Suite 300 - 990 Fort St
Victoria, BC, V8V 3K2
Canada

www.friesenpress.com

Copyright © 2021 by Levi Nathom
First Edition — 2021

Back Cover Photo – Climbers on the Face by Glen Boles

All rights reserved.

No part of this publication may be reproduced in any form, or by any means, electronic or mechanical, including photocopying, recording, or any information browsing, storage, or retrieval system, without permission in writing from FriesenPress.

ISBN
978-1-5255-8079-6 (Hardcover)
978-1-5255-8080-2 (Paperback)
978-1-5255-8081-9 (eBook)

1. BIOGRAPHY & AUTOBIOGRAPHY, ADVENTURERS & EXPLORERS

Distributed to the trade by The Ingram Book Company

Rebel in Coveralls

A Collection of Short Stories

LEVI NATHOM

For
Chelsey and David

CONTENTS

Chapter 1: Swaziland   1

Chapter 2: First Moose   11

Chapter 3: The Yukon   39

Chapter 4: Blackfish   51

Chapter 5: Camarones Café   67

Chapter 6: End of Season Whitetail   85

Chapter 7: Death Rapids   101

Chapter 8: Nakimu Caves   115

Chapter 9: Eagle   125

Chapter 10: Albert Peaks   137

Epilogue   149

CHAPTER 1

Swaziland

I grew up in Swaziland, a small African nation wedged between the Republic of South Africa and Mozambique. My father was an Anglican priest and had accepted a ministry in the parish of Manzini, one of the two major towns in Swaziland. As a result of my childhood and growing up in Africa, my life has been one of adventure. The dictionary describes adventure as *"an unusual and exciting, typically hazardous, experience or activity."* The description succinctly sums up the essence of my life.

Later in my life, when I had my own children, I would tell them bedtime stories about some of my adventures. It became so that when I went to read them a book before bed, they would say, "No, Daddy, don't read us a book, tell us one of your stories." As you can imagine, I eventually ran out of stories to tell them, so they would ask me to tell the same stories again, over and over, correcting me if I left out any of the details.

Swaziland is a Commonwealth country that gained its independence from Great Britain in 1968, while I was there. The whole country celebrated the event, for three months. I would go down to the fairgrounds in Manzini and watch the great Swazi warriors dance their war dances in commemoration, stomping their bare feet so hard on the ground that it sounded like a drum from the grandstands, and the beautiful Swazi ladies, clothed in their brightly coloured dresses, who sang and danced to their traditional tribal songs. Although at the time I did not fully appreciate the true significance of the country's independence, I was nonetheless filled with amazement and awe over the celebrations, and with the beautiful people of Swaziland.

I attended an all-white school called Sidney Williams Elementary; the local Swazis went to a different school, a Catholic school called St. Joseph's. At the time, I thought nothing of the segregation because that is how life was in Swaziland. There were rich, white settlers—generally Europeans who owned the businesses, plantations, and fancy homes—and then there were the Swazis, who were poor and lived in small grass shacks in dispersed, tiny communities throughout the countryside.

It also seemed quite normal to me, living in the southern hemisphere, that Christmas came in the middle of the summer, at the hottest time of the year. Our school year ran from February to November; we had two months off school during the summer break, in December and January. It was during the summer holidays that my father would take us on vacations to see the different surrounding countries. I suppose this is where my life of adventure originated.

Many of our vacations were into South Africa and along the beautiful Indian Ocean, where trees and vines of the jungle met white sandy beaches and monkeys ran wild and free along the beaches. On one of our holidays, though, we travelled north to Kruger National Park, which abounds with most of the animals in Africa and is among the finest game reserves on the continent. We travelled through Kruger for three days in a Volkswagen station wagon loaded with camping gear, staying at different campsites throughout the reserve.

We constantly had to stop to let many of the animals cross the road; the baboons and lions would climb on our car and we'd have to wait until they moved out of the way and off our vehicle before continuing on. My mum was terrified when the animals approached the car, especially the giant African elephants, which dwarfed our vehicle. She would yell at my dad to back up whenever we came to a herd of elephants, but he'd calmly reassure her that there was no danger and, after a while, the elephants would all cross the road and we'd carry on.

During our Kruger National Park vacation, we also travelled up and through South Africa into the countries of Southern Rhodesia and Rhodesia, now called Zimbabwe, to the great Victoria Falls, one of the Seven Wonders of the World. I remember thinking not so much about how high the great falls were but about the immense amount of water that flowed over the top of them from the mighty Zambezi River.

Sometimes in the rainy season, small islands of land, still with trees on them would break loose from the banks and float down the river and spill over the falls. We were told that some people had witnessed animals stranded on these breakaway pieces of land, like lions, that had gone over the falls.

Zimbabwe is also famous for its giant baobab trees, massive trees that looked alien, I thought as we drove past them. We visited the Zimbabwe Ruins, one of the seven man-made wonders of the world, a phenomenon similar to the Great Pyramids of Egypt.

"The Ruins" are a massive fortress, obviously built by an advanced society many years ago, made of hand-laid stones and rocks, sometimes meters thick in areas—an impenetrable fortress of intricate design. The mystery of the ruins is the provenance of the mighty rocks and stones that were used to build the massive fortress, because there aren't any of the kind for hundreds of miles around.

On our Easter break, we would travel to Mozambique, specifically to feast on prawns at a favourite seaside restaurant along the Indian Ocean. The prawns were massive. My mum and dad would order theirs "peri-peri," which were spiced with curry, I liked mine in garlic butter. On the way home we would always pick up a large tin of cashews, which are in abundance in Mozambique and relatively inexpensive.

My real-life adventures started when I became friends with my buddy Howard Emmett. I nicknamed him Spuddy and it seemed to stick. Spuddy's dad was the local pharmacist who owned a drugstore in downtown Manzini. Among the regular medicines and stock he carried in his drugstore, Spuddy's dad had a large assortment of tropical fish that he sold. Mr. Emmett, the pharmacist, was a very interesting man and I always looked forward to going to Spuddy's house for a sleepover. The Emmetts lived out of town on an acreage along a river. Spuddy's family had twelve dogs, all of different varieties, big and little. I don't remember how many cats they had but lots of them too. They also had chickens and ducks that roamed around on the lawn, some pigs, cows, a horse, turkeys, peacocks, and guinea fowl.

Mr. Emmett was a beekeeper and farmer. He also had many tropical fish tanks in which he bred and kept different varieties of fish to sell in his store. But what fascinated me the most were the homing pigeons he raised and raced.

Sometimes when I went to visit Spuddy, Mr. Emmett would take some of his prize racing pigeons and drive up the mountainous road to Mbabane with all of us kids, to release the pigeons for training. I was always amazed that, by the time we drove back home, the pigeons had already flown back and were in their cages. I would run down to check on them as soon as we drove in.

My favourite part was when Mr. Emmett would let the pigeons out to fly around for exercise. They would go straight up above the farm and fly around for hours in a big circle; we could see them from below. To get the pigeons to come down and go back into their cages, Mr. Emmett would put fresh feed out for them and then bang on the top of a metal garbage-can lid with a stick to indicate it was time to come down to eat and roost; however, on some occasions, the pigeons would fly down to the river and sit on the sandy dune banks to rest. Mr. Emmett would give Spuddy and me small firecrackers called Tom Thumbs, which he also sold in his pharmacy, and instruct us to go down to the river to scare the pigeons off the sand dunes and back into the air.

I loved going down to the river, not only to chase the pigeons into the air but to play. Spuddy and I went down every opportunity we got.

The river ran through a set of rock bluffs and then spilled down through a small waterfall made up of smooth rocks, into the river bottom below the Emmetts' house. The river bottom was all sand and water, and spread out much wider than it did through the narrow rock bluff. We would climb up through the waterfall on the smooth rocks, slide down the waterfall into the pool at the bottom, then doggy-paddle and float our way down the shallow river by kicking with our legs and walking with our hands on the sandy bottom. We spent hours in the warm water. All the dogs would come down, as well, and swim beside us.

Africa is full of dangerous animals. One of the most dangerous of all are the crocodiles, especially anywhere near water. Many a story was related of African ladies washing their clothes down by a river while their young children played in the shallows and crocodiles grabbing the children and swallowing them whole. I had seen pictures in books of crocodiles cut open with a whole child or small animal inside them. Mr. Emmett always warned us not to go down to the river alone and to always be on alert for crocodiles, but Spuddy and I never gave it much thought. We were just adventurous young boys having fun.

On one such day, Mr. Emmett had let the cows out and they had wandered down to the river to drink. The cows were standing in the river up to their knees. Spuddy and I invented a new game called swim-under-the-cows. We had fun doggy-paddling under the cows in the river; the cows didn't seem to mind us weaving in and out and around them and under their bellies. That night, when we came up to the house, Mr. Emmett told us one of the cows had its foot bitten off by a crocodile, which had made an attempt to drag the cow under the water. He looked at both of us and asked, "You kids weren't down at the river today, were you?" We both shook our heads in denial and then realized the gravity of what could have been.

On another occasion at the river, I saw a big snake wrapped around the trunk of a tree. The water was high that day, and moving fast due to a recent rainstorm; the tree was in the middle of the river. The snake must have been swept downstream from the fast-flowing river and taken refuge, clinging to the tree. I spotted it and pointed it out to Spuddy. Together we devised a brilliant plan: "Let's catch it"!

We went back up to the house and grabbed the dip-net from the swimming pool, the one with the long handle, and an empty pail. Then back down to the river we went to capture the snake. It was a big, thick snake and was fully wrapped around the trunk of the tree in three coils. We knew enough to realize this was one of the dangerous snakes of Africa. I went into the water on one side of the tree with the pail and Spuddy with net in hand approached from the other. He reached over with the net, hoping I suppose, to trap the snake against the tree inside it. I'm not sure what our brilliant plan was going to be after that, but the instant he touched the snake with the net it leapt off the tree and into the river, coming straight at me. Up until that time, I had not realized how fast a snake can swim, especially downstream with a current. At the last minute, I dove off to the side, as the snake, in a side-winding motion, swam past me and downstream, out of sight.

Since we were already down at the river with the net and pail, we thought it might be easier to catch fish with the net than snakes. As luck would have it, almost every dip of the net into the river came up with lots of little shiny fish. We proudly put them in our pail of water and carried them back up to the house to replenish Mr. Emmett's tropical fish tanks. It seemed though, that the fish from the river didn't get along very well with Mr. Emmett's tropical fish because the next morning Spuddy and I were summoned to the fish tank room by Mr. Emmett, where he pointed out all the dead fish floating on top of the aquariums. It was fairly self-explanatory how the river fish managed to get into the tropical fish tank, so we never tried that again.

The sand along the river was soft and always wet, especially when the water was low. On one of our expeditions to the river, I tried crossing a big wet patch of sand and got stuck up to my waist in the middle of it. With every movement I made, more and more water seeped in around me and I started sinking deeper and deeper into the sand. Spuddy saw I was in trouble, so told me to stop moving and said he would be right back. He found a long piece of wood that had washed up on the bank and ran back to where I was stuck in the sand. He came out as far as he could without getting stuck himself, then managed to reach me with the stick, which I happily clung onto while we both wiggled ourselves back onto more solid ground.

The scariest moment, though, happened to me one day when we were playing and swimming in the river. In one spot the river had created a whirlpool in a carved-out, sandy section along the bank, where the river made a bend. As the water swirled around, it sucked more and more of the loose sand from the riverbank into it. I remember thinking it looked like the drain from one of my baths after Mum pulled the plug.

For some reason, we thought it would be fun to dive through the whirlpool. The bank surrounding the pool was semi-circular, with the whirlpool swirling into the corner of the bank. Spuddy went first, diving from one side of the bank through the whirlpool and coming up on the other side. He had to claw his way up the sandy bank because it kept washing into the swirling current as he climbed out. But, when I went to dive in, the sand gave way under my feet just as I positioned myself, and I lost the momentum I needed to push off to dive through the whirlpool. I was immediately sucked under and into the vortex of the whirlpool. I panicked right away, trying to come up for air, but felt myself hopelessly going around and down into the vortex of the whirlpool. By sheer luck and the good grace of God, one of the dogs, a British bulldog named Cindy, jumped into the water and swam over to me. I felt her before I saw her, and managed to grab onto her collar. She carried me over to the bank where I managed to climb my way out of the whirlpool. She became my favourite dog after that.

When I was thirteen, we moved to a town called Big Bend. It was a predominantly sugarcane town, with a big sugar refinery on the outskirts. Our house at Big Bend backed onto the jungle. There was nothing but bush behind us, and I loved to go exploring and hunting. On a previous trip to Cape Town my dad had bought me a pellet gun. It wasn't very powerful, but nonetheless it was my first real gun. I hunted for birds.

The Swazis taught me how to call and hunt for birds. They would sit in the thick brush under the trees and make a sound like a flock of birds, by putting their hands over their mouth and hissing as they moved their fingers over their pursed lips. This would attract the birds to land in the trees. The Swazis all had slingshots made out of a carved piece of wood and old inner-tube rubber. They could hit a bird just about every shot; I missed with my pellet gun just about every shot. Once we had shot a couple of birds, I would follow them back to their huts where they would make a small fire and roast the birds. They would lay the birds on the hot coals, feathers and all, and rotate them by their legs until all the feathers had burnt off, then they would brush off the burnt ashes and the birds would be ready to eat. It was always an honour for me to share in the Swazis' way of life.

Further back in the bush behind us there was a large, man-made canal that supplied water to the sugarcane fields. The canal was big enough for hippos to get into. The hippos would use the canal to float down and get into the sugarcane fields. The problem for the farmers was that, once the hippos got into their fields, they destroyed more of the sugarcane crop than they actually ate, and hippos are a protected animal so there was no shooting them.

I took it upon myself to safeguard the cane fields from the hippos. I would walk along the sides of the canal, slingshot in hand and a pocketful of rocks, and on the occasion when I did see hippos, I would shoot them with my slingshot. These were the early days of my wildlife adventures.

*In April 2018, King Mswati III renamed Swaziland eSwatini, which means "land of the Swazis."

CHAPTER 2

First Moose
Part 1

I was always known as a good fisherman within my circle of friends, due in part to the fact that I had been lucky enough to catch some noteworthy giant rainbows, four of which were over twenty pounds. But it wasn't until I was in my early twenties that I started hunting. Or to be more precise, it wasn't until I was in my early twenties that I started harvesting animals; I have always been a hunter.

My best friend at the time and I fished together whenever we could—stories for another day. It was he who encouraged me to start hunting with him. He recognized my extraordinary ability to spot animals, perhaps because of my colour blindness. He did not have great eyesight; however, he was an excellent hunter and a phenomenally good shot. He had a 30-06 rifle with a fixed four-power Leupold scope. We would hunt together; I would drive and spot, he was the shooter.

His name was Keith.

Keith had shot a couple of mule deer before, but he had never shot a moose. I think it is kind of a rite of passage for any Canadian hunter to shoot a moose. At least that is how Keith and I felt about it.

Our passion was to harvest a moose. In those days, there was an open moose season, starting in September and ending in October, for bull moose. We didn't even start deer hunting until the snow was halfway down the mountains—usually the end of October.

Moose hunting was always the priority at the beginning of the hunting season, and where we hunted there were plenty of moose to support the local hunter population.

So one day Keith called me up and asked if I wanted to go moose hunting with him. He said he had a secret weapon that was sure to get us a moose. Someone had told him about a moose call, made out of a coffee tin and a shoelace. He had made one and wanted to go out and try it. Skeptical but eager, I agreed.

I picked him up early and we set off. I always drove, he always sat in the passenger side. I was happy with the arrangement and so was he. I had a three-quarter ton Ford four-by-four with a winch on the front, so my truck was the preferred vehicle.

We headed "*up the bend*", which is to say, north of Revelstoke towards Mica Creek. The "*bend*" was the mighty Columbia River, which ran north from Golden towards Revelstoke and made a *big bend* southward around Mica Creek. Prior to the establishment of Mica Creek, the area was more commonly known by its North American Indian name, Kinbasket, after one of their Chiefs.

On the ninety-mile highway that dead-ends at Mica Creek, there are three major rivers that come into the Columbia River, each with swamp-filled valleys behind them: Downie Creek at 40-mile, the Goldstream at 55-mile, and Flat Creek at 67-mile. The swamp-filled valleys only hold three types of animals: bears, both grizzly and blacks, wolves, and moose. The general area between Revelstoke and Mica Creek is also home to the woodland caribou. Back in those days, we could hunt bull caribou, but now the woodland caribou are on the Canadian endangered species list and are protected (not due to over-hunting).

Our goal for the day was to drive up the Downie Creek valley, then the Goldstream valley, then up to Mica Creek and back, hoping to see a moose. You want to shoot a moose close to the road whenever possible because anything else is a lot of work getting them out.

Keith brought along his coffee-tin moose call. I'd never seen one before, but he was convinced it was the lucky secret to getting a moose. It consisted of an empty coffee tin without a lid, and a hole drilled in the bottom. Inside the can hung the shoelace, tied with a knot through the hole in the bottom, hanging down inside the can. The idea was that, when you pulled down on the shoelace, it would reverberate through the coffee tin and make a deep, grunt-like noise, mimicking the sound of a bull moose in rut. Keith practised a few times in the truck until I told him to stop. We discovered it worked best when the shoelace was wet or damp.

We drove up the *bend,* all the side roads and all the logging roads. It was a warm, dry day, which made it difficult to see fresh tracks from the truck. We weren't having much luck—the only thing we saw was a big black bear up around 70-mile. We got to Mica Creek, turned around, and headed back to Revelstoke.

The best spot for moose is the Goldstream river and valley, at 50-mile. The Goldstream river creates a huge valley to the east behind it, made up mostly of beaver dams and swamps—excellent moose country. Logging roads had opened the valley back about fifteen miles on the south side of the Goldstream River. We had gone in as far as we could on our way up, but since there was still lots of light left in the day, I suggested to Keith that we go back up the Goldstream road again on our way back. So at 50-mile, I turned off the highway and drove up the Goldstream logging road again.

We came around a bend on the logging road at 12-mile, and standing in a little swamp beside the road on the driver's side, was a cow moose and calf, up to their bellies in water, chewing swamp grass.

"There's a moose," I casually said, like we had been seeing them all day long (funny how that happens).

I stopped and parked the truck over to the side of the logging road, away from the moose. The cow and calf both lifted their heads out of the water, looked at us, then without moving stuck their heads back into the water and carried on eating the tender swamp grass beneath it. We sat in the truck watching them.

They were close enough to the road that we could see them without needing our binoculars. We tried growing antlers on them, but it was obvious they were a cow and calf, not legal for us to harvest. Still it was nice to finally see a moose—in this case, two moose. I remember thinking the calf reminded me of a baby elephant, totally cute.

We must have sat in the truck for a good five minutes or so watching them, before Keith looked at me and said, "Do you think my moose call will work on a cow?"

"I don't know," I replied. "Give it a try."

So Keith slowly opened his door, which was on the opposite side from the moose, quietly got out, leaving his gun on the seat, grabbed the coffee tin, and went down to the ditch on his side of the road to dip the shoelace in some water. As he was doing so, I looked back at the cow and calf and suddenly noticed the bushes move behind the two of them, at the back of the swamp.

"Hey, hey," I whispered to Keith as quietly as I could. "There's something else in the back of the swamp."

He came up from the ditch and leaned into the truck. Quietly, he asked, "What?"

I said, "There is something in the back of the swamp!"

He put the coffee tin down and grabbed his rifle. He loaded it and moved behind the box of the truck. No sooner had he lifted his gun over the back of the truck than a big bull moose stuck his head out of the bushes at the back of the swamp, behind the cow and calf. He looked like one of those moose you see hanging on some hunter's wall, framed only by the surrounding vegetation.

I sat motionless in the front seat. All I could see was the bull's head and antlers. And then: BANG. It startled me, and it startled the cow and calf. They both lifted their heads out of the water, turned, and ran out of the back of the swamp and into the bushes.

The next thing I saw was the bull, in slow motion, falling forward into the swamp, splashing and kicking around like crazy, while the tiny swamp area filled up with blood, turning the disgusting green slimy swamp water into disgusting green-and-red slimy swamp water. The bull kept kicking but now was completely submerged in the water near the back of the swamp.

Then Keith came running around the back of the truck, across the road, and down to the edge of the swamp and reloaded his gun. I jumped out as he was aiming to take another shot.

"You got him, you got him!" I shouted in excitement. "He's not going anywhere."

Keith kept his gun aimed on the moose until finally it quit kicking, then floated up to the surface of the swamp. He lowered his gun and said, "Just wanted to make sure." Then he let out a big war cry: "I got him, yee-haw!!"

He unloaded his gun, put it back in the truck, came over and grabbed me and threw me up in the air. He was a big man. We were ecstatic!

"Okay, now what?" I asked.

We both just looked at each other, not knowing what to do, the moose floating in the back of the swamp.

"I'll tell you what," I broke the silence. "You shot him, I'll go get him. How does that sound?"

"How are you going to do that?" Keith asked.

"I guess I'm going to have to swim across the swamp. At least the moose floats. I'll drive the truck down as close to the swamp as I can, tie a rope around the moose's antlers, and drag it back until we can reach it with the winch. Then I'm getting out and jumping into a nice warm truck, hopefully, right?"

So that is what I did. I took all my clothes off except for my skivvies and left them in the truck, which was now parked down off the road, close to the edge of the swamp, and waded in. It was bloody cold, but worse than that it was slimy. Yuck! Oh well, I was committed.

I shoved off and started swimming across the swamp towards the moose with the rope in hand. When I got to the animal, I tied the rope around its antlers and slowly, doing a doggy paddle, side-stroke kind of swim action, began pulling the floating moose back towards the truck. I was surprised that it actually moved freely through the water, floating on top. When I got close enough to the side of the swamp where Keith was standing, I let go of the moose and, with rope in hand, climbed out of the swamp, gave the rope to Keith, and jumped in my waiting truck, which Keith had left running with the heat on.

In hindsight, other than the slime, it wasn't that bad. I quickly changed into dry clothes while Keith hooked the rope onto the winch. In no time, we had the moose out of the swamp and up to the side of the road. That was seven p.m. Saturday night. Neither of us had field dressed a moose before, let alone halved or quartered one. It took us until twelve-thirty a.m. to get the entire moose loaded into the back of the truck. I then realized a few things about dead moose. First of all, they are huge animals; secondly, they do not come equipped with handles; and lastly, two strong men cannot pick up half a moose and throw it into the back of a truck.

After a tortuous five-hour struggle, we had the moose loaded and headed for town. We went straight to the bar, looking for our friends, who were all out drinking—go figure. We went into the bar half-covered in dried blood. They were all expecting us and as soon as they saw us, they knew what had happened. Everyone came streaming out into the street to see the moose in the back of the truck. Beer and tequila shooters flowed. We were warriors.

Part 2

Sadly and tragically, in the summer of 1986, Keith drowned while fishing in the Mica Creek reservoir, which is called McNaughton Lake; some of us still call it Kinbasket Lake. I was out of town working at the time and never did hear or understand all the details of the accident. Apparently there had been three of them fishing, all good friends from Revelstoke; two had drowned, including Keith, when their boat capsized. I came home for the funeral. It was huge. Everyone from town showed up at the Catholic church. Keith and I had been inseparable as friends; everyone knew how hard Keith's death had hit me.

We were fishing and hunting partners; moreover, we were best friends, and now he was gone.

In his honour, I bought a 30-06 hunting rifle and began hunting, without him. I became a good hunter. I have natural bush instincts, something I can honestly say not all people have. I made it a mission to harvest my first moose.

I had the good fortune on one of my jobs, to work with an older man who was somewhat infamous for shooting big moose.

He had won many trophies at the Revelstoke Rod and Gun Club for biggest moose. In fact, Keith's moose came in second that year behind this older man's first-place moose. I still have Keith's second-place trophy in my collection.

One day on the job, I introduced myself to the old man and asked him what the secret of hunting moose was. He told me, "It's simple, son. You have to learn how to call them."

I asked him to show me a few calls, which he did. He said the trick to making a good call is to pretend you are bending over to go to the bathroom, cup your hands over your mouth, and let out a big grunt. Well, I practised all that summer, having no idea if I was making the moose call correctly or not.

Hunting season was approaching, and I had been lucky enough to draw a goat tag that year. In B.C., most extraordinary hunts are done by way of a draw called Limited Entry Hunting, or LEH.

Hunting for mountain goats was going to be my first priority this year for two main reasons: one, because I loved being up in the mountains; and, two, because it's best and easiest to hunt for goats before the snow or freezing rain sets in. There is nothing worse than traversing an icy side hill with a gun in hand and a heavy pack on your back.

I had a favourite mountain where I knew there were always goats; they lived there all year long. So two days prior to the opening of my draw, in early September, I set off by myself to hunt for goat. The cool thing about hunting goats—or sheep for that matter—is that they are out and visible all day long, unlike elk, moose, and deer, which are primarily nocturnal and best hunted early in the morning or late in the evening.

On this particular trip, I found myself high up in the Selkirk Mountains, on a ridge overlooking three valleys, glassing for goats. I was sitting in my camo gear, behind a small balsam tree, the only tree growing up at that height, about 9,700 feet above sea level. It was a nice, warm day, and I had been seeing lots of goats in various locations; none of them, though, were the big billy I was looking for. My goat tag was for any age and either sex, but I had no intention of shooting a nanny or a kid. The truth of the matter was, I couldn't care less if I shot a goat or not. The goat draw was just a good excuse for me to go up into the mountains and look for them.

I had just finished glassing the valley to my left, hunkered down behind the balsam, when I looked back down the ridge I was on, and to my amazement, there was a moose coming up along the top of ridge, right towards me, about a quarter of a mile away. I had never seen a moose on the very top of a mountain before; moreover, what amazed me was how it had all of a sudden just appeared there, out of nowhere, on the ridge. Nonetheless, it drew my attention immediately.

I watched it through my binoculars as it slowly lumbered up the ridge toward me. It was a nice bull, still partially in velvet. Then it dawned on me, why don't I try my moose call and see what happens. I hung my binoculars around my neck, tucked myself in behind the little tree, and bellowed out the loudest moose call I could muster, through cupped hands. Immediately, the bull stopped in his tracks and looked up toward me, straight at the tree I was hiding behind. I knew he couldn't make me out; I sat completely motionless.

There was nothing blocking his line of sight between him and me on the ridge; he just stood there staring directly at me. After a good length of time, he lowered his head and started his lumbering gait again, up toward me on the ridge. Once he started walking again, I let out another loud grunt and again he stopped and stared right at me. The thought of him running up to me had crossed my mind, but I figured I'd see how close he would come first. The second time, like the first, he stopped and stared, but this time for even longer, a good five minutes. Then he resumed his forward advancement up the ridge again. I let out a third call, but this time got a completely different reaction. The moose took a quick look toward my position, then turned to my right and started running down the slope below the ridge, toward a little alpine pond at the bottom of the slope, barking out loud like a dog, as he ran. It didn't take him long to reach the little wallow of a pond way down below the ridge, which at this time of the year was pretty much a mud hole. When he got there, he started running around the pond, stomping his feet in the soft mud and continuing to bark out his call. It seemed to me to be a challenge. He wanted me to show myself and come down there and fight him. I couldn't help but laugh to myself as the bull stomped and splashed around in the tiny mud hole. I thought: *I can call a moose!*

I never did shoot a goat that trip, but the fun of sleeping out under the stars on top of the mountain, and the images of the crazed moose are forever etched in my hunting memories.

It was now time to refocus on moose hunting. A good friend of mine, whom I had grown up with, had moved away to start his log-home-building career in Williams Lake. Gord and I had hunted together when he lived in Revelstoke, and I had been out with him many times on his trapline. Earlier that year, he had phoned me from Williams Lake and told me he'd discovered a sure moose area west of town in the Alexis Creek area, and I had committed to coming up and going hunting with him.

I drove up to Williams Lake and met up with Gord. It was great to see him again. He showed me some of the log homes in the area that he had built. It was obvious he had become a top-quality log-home builder. His homes were gorgeous.

The plan for hunting was to take my truck, because it had a winch.

"It is going to be tough getting into this spot," he told me.

Gord had discovered the moose spot by flying over it in a small plane. The only access to the area was by way of seismic lines. Seismic lines are narrow, very straight trails cut through the bush, running north and south and east and west in one-mile grid patterns. They are crude and rough, not meant for a vehicle to drive on.

We set out in the dark, early the next morning. I asked Gord as we were driving what was considered a legal bull in this particular hunting area. He told me, "Any male moose."

As he put it, "If you can see balls, shoot it!"

We turned off Highway 97 at Williams Lake and headed west toward Alexis Creek; the highway eventually ends at Bella Coola on the Pacific coast. The weather was just starting to turn; you could see your breath in the air. Winter was not far off. As we drove out past one of the many side roads, I noticed a herd of cows pressed up against a fence, which was preventing them from coming onto the highway. They looked disturbed. Their eyes were wide open, and steam was coming off them. I stopped and turned the headlights of the truck on them. You could see steam coming out their nostrils in the darkness of the morning as they breathed heavily.

"What's with that?" I asked Gord.

"Free-range cows," he replied. "Probably chased down by a pack wolf. Lots of wolves around here."

I turned the truck back to the highway and carried on.

The sun came up and it was going to be a nice warm, clear day. An hour or so down the road, Gord yelled out, "Stop, stop! Look, right there, it's a moose!"

I swung the truck over to the side of the highway, stopped, and got out with my gun on the driver's side, opposite to the moose, which was standing up on the bank on the high side of the road behind a fence. I peered over the side of the box of my truck and looked through my scope. Sure enough, it was a moose. I could tell it was a young one but there were no antlers. I looked back inside the truck where Gord was sitting and quietly whispered, "It's a cow. No antlers."

No sooner had I looked back over toward the moose than it jumped over the fence and started coming down the bank toward us, heading for the other side of the highway. I could see there was something definitely wrong with it. The moose's back hind right leg was broken, dangling like a loose appendage. The leg simply dragged over the fence as the moose jumped it. The moose was in obvious trouble as it hobbled its way down to the highway.

"Shoot it," Gord whispered out to me, still sitting in the passenger seat. He could see the moose was injured as well.

"I can't see antlers," I whispered back.

"Shoot it anyway," he said. "It's going to die."

The moose came down the slope, through the ditch and up onto the highway less than fifty yards behind the truck. It couldn't care less that I was standing right there, outside of the truck with a gun. Remembering what he'd said about seeing balls, I looked underneath the moose but couldn't make out a set of testicles, so chose not to shoot. The moose crossed the highway and jumped over the fence on the low side of the road. I followed it down through the trees on the opposite side for a while, thinking I should shoot it to take it out of its misery, but somehow couldn't pull the trigger. After a few hundred yards, I turned back and returned to the truck.

"Why didn't you shoot it?" Gord asked in a scolding manner.

"It wasn't legal," I said, defending myself.

"Well, it's wolf meat now," he said.

I got back in the truck and started driving again, wondering if I'd done the right thing, wondering how the moose had become injured, and if it was already being chased by wolves. I guess I will never know.

When we arrived at Alexis Creek, we pulled in and filled up with fuel. Then we turned north off the Bella Coola highway and onto a gravel logging road. After a couple of miles, Gord directed me to turn back east again onto a small seismic trail. He was right, it was going to be rough and slow going.

On portions of the seismic line, it was quite obvious that a forest fire had gone through years ago, and, where it had, the newly regenerated forest that had grown back in was thick and dense with pecker-pole pine trees. The seismic line had been cut only wide enough through the dense, small regen forest to get a tracked vehicle through. The trail was barely wide enough to fit the truck, trees scraping the mirrors on both sides.

At one point, we came to a sheer cliff, more or less a complete drop-off. Other hunters before us had cut and cleared a truck path around and down past the rocky drop-off of the seismic line. I drove down the precarious detour, praying it would come out on the bottom somewhere; there was going to be no way of turning around. We managed to get to the bottom of the hill, circumventing the cliff and returning back to the rocky and bumpy seismic line again, only to find we were challenged with another obstacle. At the bottom of the hill was a swamp created by beaver dams. I looked at Gord with a look that needed no explanation.

"I told you it would be tough getting in," he said, averting my stare.

The prior hunters had fashioned a corduroy type of bridge across the creek in the swamp, laying down a bunch of logs over the wettest parts. The wood was old, and I knew it would collapse under the weight of the truck if I drove over it, but we had few options. Turning around and going back was not one of them.

We pulled the winch line out and hooked it around one of the trees beyond the corduroy bridge, on the far side of the swamp. Then I slowly drove over the logs, winching my way across at the same time. The logs did break under the truck, but somehow we managed to winch our way across the wettest part of the swamp and around the beaver dam at the end, and eventually got ourselves back onto the seismic line again, which was more or less clear sailing on solid ground from there on. It took us all day to work our way to where we were going. Just after it got dark, we came to an intersection of two lines, each running in a different direction, much like a cross.

"We are close now; turn down this line, and it should only be a mile or so from here," Gord directed me.

I did as instructed, and it wasn't long before we came to a creek that crossed our trail, in a nice open meadow area.

"We'll make camp here," Gord said.

In the dark, we set up our wall tent in a grassy area among the lodgepole pine trees and built a fire. Then it was time to relax and have a few drinks before going ten toes up!

We got up early in the morning the next day. It was still pitch black out. There was an eerie fog that had rolled in through the trees that morning, making it even harder for me to distinguish my new surroundings.

Gord came over to me, and in a sedate and quiet voice said, "The moose could be anywhere. Keep really quiet. You cross the creek and go down that line and I'll walk back to the junction and head up the other way. We'll meet back here for breakfast in a few hours and figure it out from there."

I nodded in agreement, grabbed my pack and rifle, and was just about to head out when, off in the distance, we heard wolves howling. We both stood there in the dark and fog of the morning, listening to them. The individual sounds of the wolves started becoming more collective and louder—the pack was moving in our direction. It was an ominous feeling, standing in the dark fog, listening to a wolf pack travelling through the bush, obviously on the hunt. Eventually, they came so close to where we were camped that we could make out the individual sounds of each wolf as they invisibly passed by us. The hair on my arms stood up. It was a chilling moment.

Gord looked over at me. "Pretty cool, huh?"

After the pack had passed and we could no longer hear them, I motioned that I was heading off. I crossed the little creek near the camp and walked down the seismic line heading north.

I always carry a compass with me in my pack. This kind of country was unlike the kind of mountainous country I was used to hunting. At least in the mountains you can always head down if you get lost. Here, it was undulating, tree-covered hills with no peaks. It would be quite easy for me to get lost or turned around if I didn't pay attention to my compass.

Slowly, the fog dissipated, and the morning sun chased off the darkness. It was a pleasant day, and I was brim full of anticipation. I'd only gone a few hundred feet down the trail when I came to a tree that had an old white moose antler stuck up in the branches. It was obviously an old *shed*, an antler from a winter moose, but what struck me the most about it was how high up in the tree it was lodged. There was no way I could reach it from the ground. I estimated it to be about twelve feet up in the tree. *Impressive*, I thought. *A big moose*. I walked on past it.

I didn't particularly care for walking along the seismic line because it was fairly open, going through dried-up marshes and open meadow areas. I felt exposed and easily visible. The surrounding trees were all lodgepole pines—a mature stand, easy enough to see into and open enough to walk through, so I stepped off the seismic line and into the trees. As I looked further through the trees, I could see an opening up ahead. It looked like another meadow area, so I walked through the trees, over to it.

It was a small, grassy meadow in the middle of the pine forest, about the size of a football field. When I got to the edge of it, I glassed it for signs of moose but there was nothing. I noticed beyond it what looked like another open meadow, so I worked my way through the trees around the first meadow and came out into a much larger meadow.

I crawled out through the tall, dried grass of the meadow on my hands and knees so as not to be seen, over to a small clump of trees, just off the edge of the big meadow. There, I quietly removed my pack and gun and got my binoculars out. It was a long, narrow meadow, and as soon as I looked through my binoculars I saw something at the far end.

At first, I wasn't sure what it was. It looked like the head of a cow moose, but it was completely grey. *Like a small elephant,* I couldn't help thinking. Back home, most of the moose are either brown-and-black or black.

I stared and stared at it for a while, realizing finally that it was a cow moose, but all I could see was its head and neck.

It was obvious it didn't know I was there; it was motionless, looking straight ahead into the bushes in front of it, at a ninety-degree angle away from me, in the middle of the high grass near the back of the meadow. It looked strange to me, I couldn't figure out exactly why.

I re-positioned myself under the clump of trees and watched it for a while. It remained there, motionless, looking straight ahead. Then I thought to myself: *It is not going to come any closer, and I can't step out into the open to get closer to it without being seen, so, hey, why don't I try my moose call?*

I rested my binoculars on my camo pack, cupped my hands, thought about what the old man had told me, and let out a mighty grunt. Immediately, a big bull broke out of the bushes in the exact spot the cow had been staring at. The bull ran out into the open meadow straight toward me, snorting and grunting.

Holy crap, I thought. *He's going to run right over me.* I slowly placed my gun over my pack, released the safety, and took aim at the big bull. He wasn't stopping. He was running straight for the clump of trees under which I lay.

Closer and closer he came through the open meadow. With each step, he got bigger and bigger in my scope. Finally, and fortunately, he came to within about seventy-five yards of me, stopped, looked at the clump of trees I was hiding in, then turned broad side to me. I couldn't believe my luck.

Boom, I fired off a round. As soon as I did, I realized what was happening with the cow in the background. She had been bedded down in the tall meadow grass. When she jumped up at the sound of the gunshot, I realized I was much closer to her than I had thought. I had somehow thought she was standing up, way off in the tall grass at the back of the meadow.

She was a monster cow moose, bigger than any I'd ever seen before, and up until now I hadn't realized either that there was another cow bedded down beside and behind her. They were both giant grey cows. Like I said, they reminded me of two small elephants.

Off they ran into the trees at the back of the meadow.

I looked back at the bull, still standing broadside, looking at me. I reloaded and fired again. This time in a split second through my scope, I saw him stumble forward and disappear into the bushes on the side of the meadow. It's amazing how the brain can analyze something the eye only sees for a micro-second. I knew I had hit him.

Remain calm, I said to myself while my heart was racing a mile a minute. *Don't chase it. You got it! Wait, be patient.*

All these thoughts ran through my head. Everyone had always told me not to panic when you shoot an animal, to give it ten or fifteen minutes, then go after it. The worst thing you can do is chase a wounded animal. Let it go off and die.

To hell with that I thought. I jumped up, grabbed my pack and binoculars, reloaded my gun, and walked quickly around the edge of the meadow to where the bull had gone in. As I got to the spot, I realized the meadow actually swung back around and opened up, and there lying on the ground, right out in the open meadow, was the big bull, breathing his last breaths. Without hesitation, I walked up to him and fired a final kill shot to his head. WOW. My first moose, and I had called him in. I was tremendously proud of myself; I had done everything perfectly.

I set my gun down, took off my pack, and undressed a few layers of clothes. I had two sharp knives in my pack for just such an occasion. I dug through my pack for a few short pieces of rope, which I used to tie around the back leg of the big bull. I lifted his leg up as far as I could and tied it off to a tree nearby. This exposed his belly so I could start field-dressing him. I did the same with the front leg, tying it off to another tree, in effect spreading the moose open for easier access to cleaning it. It didn't take me long to field-dress the moose; my previous experience with Keith's moose helped. I was careful to save the heart and liver. I left the heart with the moose for now, but wrapped up the liver in a cotton sheet, put it in my pack, and headed back to camp. I wasn't sure what time it was but it felt like I had been gone a long time.

When I got back to camp, Gord hadn't returned. So I build a fire, put a coffee pot on to perk, dug through the cooler for some bacon and eggs, and started to cook breakfast. Once the bacon started cooking, I removed the liver from my pack and sliced off a few thin pieces. No sooner had I started cooking breakfast than I heard Gord coming down the seismic line. He walked into camp and sat down at the picnic table we had fashioned.

"Just in time for breakfast," I said. "See anything?"

"Lots of sign but no moose," he replied. "How about you? I thought I heard gunshots over your way?" Gord pointed in the general direction of where I had shot the moose. "Did you hear them?"

"You did, I did," I said. "It was me."

"What were you shooting at?" he asked.

"What do you think I was shooting at, a moose!"

"What do you mean?" he said.

"I shot one," I replied.

"Get outta here," he said in disbelief. "No, you didn't."

"Look in the frying pan," I said. "What does that look like to you? Fresh moose liver, perhaps?"

He looked at me, and then at the fry pan with the fresh moose liver sizzling in the bacon fat, then at the fresh slab of liver on the table behind me, and realized I wasn't pulling his leg. He grinned and smiled.

We both sat down and enjoyed a very fine, fresh moose-liver breakfast. It was a good morning indeed.

Part 3

One of the beneficial consequences of Keith's death was that I befriended Bruce. Bruce and Keith had grown up and gone to school together in the B.C. lower mainland. Keith had moved to Revelstoke to be a logger; Bruce had stayed in Vancouver to finish his education. I met Bruce through Keith. Bruce often came up to go fishing with Keith and me. Now, with Keith's passing, we had made a pact to be hunting partners. Bruce is a tall, naturally gifted athlete who went through university on a basketball scholarship. I think, however, it would be fair to say that when we first started hunting together Bruce was a little green. But his fitness level and commitment could not be denied, so we made a good partnership. As it turned out, many years passed before Bruce ever shot his first moose, but when he did it was a memorable one.

By now, hunting moose up the "bend" was by LEH hunt only, and the odds of getting a draw for the Goldstream were ninety to one. We both put in for a moose draw every year, without success, until one year.

Bruce called me up in the spring and said, "Guess what, I got a bull draw for the Goldstream."

Bruce took a week off from his job in Vancouver, and up the bend for moose we went.

I had a travel trailer, which was to serve as our base camp for the week. We arrived a day earlier than the opening of the draw, so we could scout out the best areas. We figured we'd get a moose on opening day and wanted to be ready.

Well, best-laid plans of mice and men don't always work out, and we found ourselves on the fifth and last day of our hunting trip without a moose. We had seen a few cows, but that was all. The weather was unseasonably warm, and there was nothing moving, very little sign anywhere. To be honest, I was fed up and bored with the hunt. We had driven every logging road north of Revelstoke to Mica Creek and back, and I was tired of driving and seeing nothing.

On our fifth morning of the moose hunt, I said to Bruce, "I'm sorry, bud, but if we don't see anything today by noon, we're packing up and heading home." Bruce nodded. He was fed up too.

"Where do you want to go today?" I asked as we finished breakfast.

"I don't care, anywhere," Bruce said in resignation.

We decided to go back up French Creek again, which was on the north side of the Goldstream River. By now, the labyrinth of logging roads up the Goldstream valley was extensive. The new dam at Revelstoke had changed everything. There were even two five-star CMH heli-skiing lodges in the Goldstream valley now, the Adamants and the Gothics. There was constant traffic on all the back roads. We drove to the end of the French Creek road, looking into the swamps, there and back again without success.

"Let's go home," I said. "I'm done!"

"Yeah, sure. I'm done, too," Bruce replied.

I drove the few miles back to where we were camped on the Goldstream road. When we got back I said to Bruce, "We agreed to hunt until noon today, right? So let's go down the Goldstream road one more time. It's still early." I didn't want to give up on Bruce's moose tag, for his sake.

"Sure, whatever," he said. So we drove down the Goldstream road one last time, slowly, without hope or expectation. It was mid-morning—little hope of seeing anything at that time of the day. Amazingly though, we came around a corner and off to our left, standing in the back of a swamp, was a cow moose. I stopped the truck and parked on the road. We both watched her for a while; she wasn't going anywhere. After a while, Bruce said, "I wonder if I could get around the back of that swamp."

"What for?" I asked.

"I don't know, just to check it out, to see if there are any other moose back there."

"Go ahead," I said. "I'm not going anywhere. How are you going to get around the back of the swamp anyway? It's way too wet and thick in there."

"I don't know. I'll walk down the road and see if there is an opening somewhere."

So out of the truck he got and walked down the road, along the side of the swamp.

I looked back at the cow, still standing in the swamp eating swamp grass. I picked up my binoculars again and started glassing the thick bush behind her. As I scanned the bushes behind the swamp, from left to right, I saw what I knew instantly was an antler. I stopped and focused in on the spot. It took me awhile to find it again, among all the alders and willows, but I just knew in my heart of hearts I'd seen the glint of an antler. I looked again and, sure enough, found it. At first, I couldn't make out what I was seeing, but then the antler moved and I saw the sun reflect off it. I looked again and could barely make out what was a big bull moose bedded down in the bushes in the back of the swamp. I jumped out of the truck and waved my arms at Bruce, who was a considerable way down the road by now. He looked back at me and shrugged his shoulders as if to say, "What's up?" I frantically waved for him to come back.

He turned around and started coming back up the road toward the truck. I glassed back into the swamp again, and just as I did the big bull stood up. He was still hidden from sight by the alders and willows but I could see him now, the whole outline of him, whereas before when he was bedded down I couldn't make him out properly.

I looked back at Bruce and frantically waved my arms again for him to come back. He walked a little quicker. I motioned for him to run. He must have understood that something was up, so he broke into a slow jog back to the truck. I grabbed my gun and loaded a round into the chamber. I got out, leaned over the back of the truck and picked up the moose in my scope. The bull was walking through the brush, from my right to left, over towards where the cow was. I had a clear shot at him. I looked back at Bruce. He was getting closer, but the bull was about to walk away, back into the bushes.

I waved frantically again at Bruce, watching the big bull through my scope the whole time. The bull was disappearing behind some thick brush, again I had the chance to shoot him broadside but didn't. Finally, Bruce came up to me beside the truck. "What, what?" he asked.

"There's a bull in there," I said to him. "You see those two trees way back there? He's going to cross in between them. Aim right between those two trees. He's going to step out any second."

Bruce brought his gun up over the back of the truck and took aim. I don't think he'd even seen the bull yet, but no sooner had he taken up a shooting position than the bull stepped out into the open between the two trees, and BANG.

I was watching everything through my scope. Down went the bull, and it never moved again. "You got him, you got him!" I said, all excited.

Bruce hardly had any time to think, but he had made a perfect head shot. I could see the bull's antlers sticking up as it lay dead in the bushes at the back of the swamp.

"Good job," I said, and high-fived him.

"Thank you for waiting and letting me shoot it," he said, gratefully.

I told him I could have shot it over and over again, and near the end I almost did pull the trigger because I wasn't sure he was going to make it back in time. But it had all worked out for the best.

"So now what?" he asked

"Well, since you can't swim, I guess I'm going to have to go get it," I replied. "I'm going to need a saw to cut a trail through the brush, and the long rope."

My moose-hunting equipment had improved over the years. I had a special toolbox that I always took with me moose hunting. Among other things, it held a snatch block and come-along. I told Bruce to hook the snatch block up as high as he could on a tree beside the road. I grabbed my Swede saw and the long rope, walked down the bank, and waded into the swamp. I swam across to the back of it, to where the moose was lying. When I got to it, I was impressed.

It was a nice bull, lying in a somewhat convenient position, with its head pointed toward the swamp and truck. I cut out a path through the alders down to the swamp, tied the rope around the antlers of the moose, and waved over to Bruce that I was all hooked up. He ran his end of the rope through the snatch block and tied it to the back of the truck. He drove slowly down the road at a ninety-degree angle to the snatch block and slowly, we dragged the moose down and into the water at the back of the swamp.

Once the moose was in the water and floating, Bruce stopped the truck and backed up. He re-tied the rope directly to the back of the truck and pulled the moose across the swamp, with me hanging on. Once across the swamp, it was just a matter of pulling it up the bank to the road, which we did by shortening the rope and pulling the moose up the bank with the truck. Once we got the moose up and onto the road, Bruce started cleaning it and I went and changed into some dry clothes. One of the things I like best about hunting with Bruce is the way he field-dresses an animal and cares for the meat. He is surgeon-like cleaning any animal. That makes all the difference when it comes to eating it.

As we were cleaning the moose on the side of the road, a logger I knew from town pulled up and asked us if we needed a hand getting the moose in the back of the truck. Again, my moose-fetching skills had improved over the years; I had two two-by-fours in the back of the truck and a sheet of plywood. After cutting the moose in half, the three of us easily slid it up into the back of the truck. We thanked him, and off he went.

I grabbed a couple of beers out of the cooler in the truck and handed one to Bruce. "You want to know something?" I said as I turned to him and raised my beer to make a toast.

"What's that?" he said.

"This is the exact same swamp where Keith shot his first moose, and it was exactly twenty years ago to the day."

In loving memory of Keith Hiebert, one in a million.

CHAPTER 3

The Yukon

The Spell of the Yukon
by Robert Service

I wanted the gold, and I sought it;
I scrabbled and mucked like a slave.
Was it famine or scurvy—I fought it;
I hurled my youth into a grave.
I wanted the gold, and I got it—
Came out with a fortune last fall,—
Yet somehow life's not what I thought it,
And somehow the gold isn't all.

No! There's the land. (Have you seen it?)
It's the cussedest land that I know,
From the big, dizzy mountains that screen it

To the deep, deathlike valleys below.
Some say God was tired when He made it;
Some say it's a fine land to shun;
Maybe; but there's some as would trade it
For no land on earth—and I'm one.

You come to get rich (damned good reason);
You feel like an exile at first;
You hate it like hell for a season,
And then you are worse than the worst.
It grips you like some kinds of sinning;
It twists you from foe to a friend;
It seems it's been since the beginning;
It seems it will be to the end.
I've stood in some mighty-mouthed hollow
That's plumb-full of hush to the brim;
I've watched the big, husky sun wallow
In crimson and gold, and grow dim,
Till the moon set the pearly peaks gleaming,
And the stars tumbled out, neck and crop;
And I've thought that I surely was dreaming,
With the peace o' the world piled on top.

The summer—no sweeter was ever;
The sunshiny woods all athrill;
The grayling aleap in the river,
The bighorn asleep on the hill.
The strong life that never knows harness;
The wilds where the caribou call;
The freshness, the freedom, the farness—

O God! how I'm stuck on it all.

The winter! the brightness that blinds you,
The white land locked tight as a drum,
The cold fear that follows and finds you,
The silence that bludgeons you dumb.
The snows that are older than history,
The woods where the weird shadows slant;
The stillness, the moonlight, the mystery,
I've bade 'em good-by—but I can't.
There's a land where the mountains are nameless,
And the rivers all run God knows where;
There are lives that are erring and aimless,
And deaths that just hang by a hair;
There are hardships that nobody reckons;
There are valleys unpeopled and still;
There's a land—oh, it beckons and beckons,
And I want to go back—and I will.

It's the great, big, broad land 'way up yonder,
It's the forests where silence has lease;
It's the beauty that thrills me with wonder,
It's the stillness that fills me with peace.

If you have never been to the Yukon then you have never really seen Canada. It's hard to describe; there is a spirit there that grips your soul. You have to be there to understand. Vast is an understatement to describe the valleys and mountain ranges. As far as you can see—and in some places in the Yukon, you seem to be able to see to the edge of the earth—there is nothing but trees, valleys, and mountains. It's overwhelming.

If you are fortunate enough to go to the Yukon, then a trip up the Dempster Highway is a must. The Dempster Highway runs in a north-easterly direction, starting twenty-five miles west of Dawson City and ending in the Arctic at Tuktoyaktuk. The Dempster crosses the Peel River and the Mackenzie River delta and runs through the Ogilvie and Richardson mountain ranges, but the most notable feature of the Dempster is that it is the only highway on earth where you can drive across the Arctic Circle.

One of the first things I did when I arrived in the Yukon was to pick up a copy of the hunting regulations, and a copy of the regulations pertaining to subsistence living—not that I intended to do either. It was just interesting to read about how to apply for free land, if you intend to live off the land, and all the requirements that go along with subsistence living.

But I had to laugh when I read the hunting regulations. In bold print on the first page of the hunting regulations, it declared that it was unlawful to leave your caribou gut piles in the middle of the highway because they freeze and cause hazardous bumps on the road.

The Yukon is home to the Porcupine caribou herd, a herd of 120,000 animals, the largest herd of animals in the world. As you drive up the Dempster, it is common to see herds of caribou out on the tundra, grazing. It is also common to see the First Nations hunters camped along the highway, waiting for the caribou to cross the road, or to at least come within shooting distance of the road.

On one occasion, I stopped and watched the hunters as they were parked along the highway. They had a convoy of pick-up trucks, each with a couple of hunters in the back. One hunter was set up for spotting, with a tripod and spotting scope; the other was the shooter, with a tripod and a high-powered rifle. As the distant caribou herds moved, so did the hunters. While they waited for the caribou to graze closer towards the highway, yellow station-wagon taxi cabs from the local First Nations band kept coming alongside, loaded with provisions for the hunters, as well as bringing out replacement hunters to spell off those in the back of the pick-up trucks.

It was unlike any kind of hunting I had ever observed. After a while, I thought it best to move along and leave them to it.

On my way back down the Dempster, there was no sign of the caribou or the hunters; however, at various locations along the highway, there were bright blue tarps spread out. My curiosity got the better of me, so, looking around to make sure no one was watching, I pulled over and walked over to one of the tarps. Neatly arranged on them were different body parts of the caribou, drying in the sun, obviously from a successful harvest. The animal parts had been expertly butchered and carefully placed in such a way that made me think each tarp was dedicated to certain families or members of the band, although I could not be sure if that was their intention. It was very tempting to take one of the caribou tenderloins that had been perfectly filleted and left to dry on the tarp, but my respect for the First Nation hunters was greater than my desire for the meat, so I left everything exactly as I found it.

The valleys along the Dempster are unending; grizzly bears graze the plains like cattle in a field. I saw more grizzly bears than moose, even though they claim in the Yukon there are more moose than people.

You will want to drive all night to witness the northern lights in their full, spectacular spectrum, and to discover that it never really gets dark, merely different shades of pink, purple, red, and blue.

Once you get past the Arctic Circle you start seeing snowy owls and Arctic foxes. The highway is marked with high posts on either side so that, during white-out snowstorm conditions in the winter, you can make out where the highway goes. On almost every one of the posts over a hundred-mile stretch, there sat a snowy owl. I surmised it must be an area with a healthy rodent population, to support that many owls and foxes.

Amazing as it may seems, the Dempster Highway is a go-to vacation spot for Japanese tourists. The notion of being able to cross the Arctic Circle attracts Japanese adventure seekers every year. Most of them ride the Dempster on bicycles just to be able to say they rode across the Arctic Circle. There is even a story of a fifty-six-year-old Japanese man who came over in the winter and rode his bicycle across the Arctic Circle in sub-zero weather. He returned home to Japan as somewhat of a national hero, so the story goes.

I have witnessed the trains of bike riders on my trips up the Dempster, even once befriending a Kiwi bike rider whose lifetime ambition it was to ride across the Arctic Circle.

Bruce was his name. I found him sleeping in a pup tent, in a freezing cold wind storm at Eagle Plains, and offered him a ride the rest of the way up the Dempster to the Arctic Circle, but he would not accept. He had flown over from New Zealand, landed in Vancouver, and ridden his bike all the way up to the Yukon in order to ride across the Arctic Circle. He was not going to let me deny him his moment of glory, fifty miles away from his destination. He did agree though, to accept a short lift from me on my way back. So after I had driven all the way up to Inuvik and back, and Bruce had ridden up and across the sixty-sixth parallel, I met up with him again on my way back and gave him and his bike a lift back down the Dempster. But after only a few hours or so, he couldn't stand the notion that he might be cheating on himself by not riding his bike, so he asked me to let him out. I kept in touch with Bruce for many years after, reading his postcards from all over the world, where he was still riding his bike.

It was Bruce who drew my attention to the international desire to ride across the Arctic Circle, especially those in Japan. It was Bruce, too, who related one of the most bizarre stories I have ever heard.

None of my stories start with *"once upon a time,"* so—once upon a time, a Japanese adventurer decided to come to Canada, to ride his bicycle across the sixty-sixth parallel, the demarcation line known in the northern hemisphere as the Arctic Circle.

The Japanese man had heard about the Yukon from one of the many travel shows he'd seen on television. Tourism is huge in Japan, and so are travel shows. The Japanese man had an adventurous heart and a courageous spirit.

He flew into Vancouver, then caught a connecting flight to Whitehorse, the capital of the Yukon Territory. From there, he boarded a bus to Dawson City. At Dawson City, he bought the appropriate provisions for his two-week ride across the Arctic Circle, much like the hopeful prospectors of a century earlier.

His first day took him to the junction of the Dempster and the Klondike highways, about twenty-five miles west of Dawson City. There, he turned northward and began the gruelling uphill ride toward the Arctic Circle. Some of the Dempster Highway is paved, some is gravelled with sharp black rock from the Ogilvie Mountains, and the rest is made up of dirt, dust, pot holes, mud, and any combination of those, depending on Mother Nature's attitude of the day.

There is always a certain amount of trepidation when one is travelling in a foreign land, especially when fellow travellers are infrequent and solitude is frequent. After a few days of gruelling riding, the Japanese adventurer started imagining that he wasn't alone, although it had been days since he had actually spoken to anyone. As he rode his bike, he couldn't help but get the feeling that he was being watched.

One night, after setting up his camp and cooking his supper over an open fire on the side of the highway, as he had done for the past few nights, the feeling of being watched was so overwhelming that he purposefully looked around his camp and into the trees to assure himself that he wasn't going mad. Much to his dismay and discomfort, way back in the trees he saw a pair of eyes staring at him. That night, he pulled his tent closer to the fire and had a restless sleep, waking up continually and peering outside his tent.

The next morning he scouted the area around where he'd made camp and saw nothing. So, back on his bike, he climbed and continued his gruelling trek. After pedalling hard for a few hours, he began again to get the eerie feeling that something or someone was watching him. He tried to ignore it and pushed on.

That night, at his campfire, he was overwhelmed with the feeling that he was being watched again. He got up and had another look around, and, sure enough, he caught a glimpse of what appeared to be the same set of eyes he had seen the previous night, staring back at him. But this time they appeared to be quite a bit closer.

Once again an unnerving night was had. In the morning, he began riding again and, after a while, had the same feeling he was being watched. At the end of the day, he made camp again and waited for the eyes to appear in the trees. When he spotted the eyes this time, they were much closer, and it was only then that he could make out the silhouette that belonged to them. It was a giant dog who, this time, wasn't retreating into the bush as it had on the previous encounters. So the quick-thinking Japanese adventurer went back to his campfire and picked up a couple of bones he had discarded, went back toward the giant dog, and threw them into the bush in hopes of getting rid of his unwanted companion.

The next morning he got back on his bicycle and started riding again. As he rode, he looked over his shoulder into the bush along the side of the highway, and every once in a while caught a glimpse of the giant dog. That night again, sitting around the campfire he saw the giant dog, now more apparent than ever, standing at the edge of the glow of the fire light. After the man had finished his supper, he once again threw his dinner scraps over to the giant dog. The dog ate the scraps and left.

The Japanese adventurer set off the next day, but this time the giant dog did not try to hide himself. He ran along beside the Japanese adventurer, some distance off in the bush. Or perhaps the man had become more accustomed to seeing him more easily in the bush. Nonetheless, the giant dog was obviously following him.

That night, the giant dog came into the camp and sat beside the fire, waiting to be fed. The charitable traveller fed the giant dog, who then wandered a little ways off and lay down in the bushes at the edge of the campsite.

After a couple of days of this, the giant dog came right out onto the highway and ran behind the adventurer as he pedalled as fast as he could up the Dempster Highway.

Again that night, the giant dog appeared at the campfire for his supper. It was then that it dawned on the Japanese traveller that he was desperately running out of food. That night, he did not have enough food for both himself and the giant dog. He knew Eagle Plains was only a day's ride away, where he could stock up on more provisions, but for that night, there simply wasn't enough food for two. So he cooked up the last of his rations and gave them to the giant dog.

A young couple with their children in the car were coming back from Fort McPherson on their way into Dawson City. They had stopped at Eagle Plains to fuel up and let the kids out to stretch their legs. A few miles later, they came around a corner, and standing in the middle of the road, was a Japanese bicycle rider, waving his arms in the air frantically trying to get the couple to stop. As soon as they did, the man opened one of the car doors and jumped in, hysterically yelling and begging the couple to give him a ride.

In broken English he related his tale of the giant dog to them and how he had run out of food and thought he was going to be the giant dog's next meal. The Japanese adventurer graciously accepted a ride back to Dawson City and never did make it across the Arctic Circle.

CHAPTER 4

Blackfish

I've spent a lot of time on the water off the west coast of British Columbia, fishing, boating, and exploring.

Like most, seeing wildlife in the ocean is exciting, an inherent privilege of being on the Pacific coast. I have always wondered why there are so many different species of wildlife on the earth. Why not just one type of whale or one type of butterfly?

Anyway, let me tell you about an amazing fishing trip I went on. It was up in Prince Rupert. I was coming back from one of my Yukon trips, with my truck and camper. I had taken the ferry from Skagway down to Juneau, then out to Sitka to visit relatives, and then back down the Inside Passage to Ketchikan, Wrangell, and St. Petersburg, before reaching my destination of Prince Rupert. If you have never been to southeast Alaska, it is also a must see!

I still wonder how southeast Alaska is not Canada.

There is no difference between the two geographically, all of it is stunningly beautiful. But that is a whole different story about how Sam Steele of the North West Mounted Police drew a line in the sand so to speak, above the Chilkoot Pass at Bennett Lake, and declared the territory beyond that Canada, to deter prospectors from entering into Canada and starving to death without the necessary provisions, during the 1898 gold rush. Somehow the Americans managed to claim the bottom of the southeast panhandle.

The area above Prince Rupert, and for that matter the entire west coast of British Columbia and Alaska, is pristine and untouched, abundant with wildlife and biodiversity.

Humpback whales, minke whales, and grey whales travel up the west coast from as far away as Mexico and the Hawaiian Islands to feed on the nutrient-rich waters of the Pacific Northwest. Sea otters are a common sight, floating on their backs among the kelp beds, eating clams and abalone, their numbers back from the brink of extinction. (It's hard to imagine a world without sea otters because we might have killed them all, for fur!)

There are puffins and all types of different sea birds. Colonies of seals and sea lions. Porpoises that skate along the wake of your bow. But nothing seems to capture our attention or imagination more than seeing a pod of orcas or, as some call them, killer whales. My friend Captain Ben, who I was about to catch up with in Prince Rupert and go fishing with, affectionately calls them "blackfish."

Orcas, which are from the porpoise family and aren't whales at all, are the top predators in the ocean. Even great white sharks fear the incredibly intelligent blackfish.

Most of the Alaskan and Canadian Pacific coastline is protected by outer islands and land masses, keeping the inner waters calm and rich with food. Thousands of islands from Seattle to Anchorage make up this incredible and diverse archipelago of nature.

Southeast Alaska becomes Canada at the Portland Channel, which takes you into Stewart, B.C., and Hyder, Alaska. If you drive by road to Stewart, which I have done on several occasions, and then cross over into Hyder, and then keep driving, you'll find yourself back in B.C. again. I remember once when I was up hunting goats in the area, I came across a sign nailed to a tree way up in the mountains that stated I was now entering the state of Alaska. I wondered if the goats knew which country they were in.

If you do happen to go to, or through, Hyder, then a must stop is at the Hyder saloon to get *Hyderized*. They say you are not a true Alaskan if you have not been *Hyderized*.

Walk up to the barkeep in the Hyder saloon and tell him you are new to Alaska, and that you want to get *Hyderized*. He will pour you a generous shot of moonshine—150-proof alcohol—and serve it to you in a shot glass with a frozen toe in it. You must shoot the glass back until the toe hits your lips. Only then have you become *Hyderized* and can stay for beers with the locals.

On the way down through the Inside Passage, the ferry stops at Wrangell Island. At Wrangell, a stroll on the beaches is a must while you wait for the ferry to take on passengers and provisions. The beach is full of dark-red, perfectly shaped amethyst stones—like rubies, but nowhere near as rare of course. The whole beach is full of these bright gemstones. You can't miss seeing them, so I stopped to collect a few, much like looking for seashells.

My plan was to get into Prince Rupert and give Captain Ben a call upon arrival. Ben was an incredibly accomplished "salty dog," for lack of a better description. He had been running boats up and down the west coast all his life and knew his bearings—weather patterns, tides, and currents—better than anyone. Word had it that a young couple travelling the west coast in a sailboat had become panic stricken and were dead in the water somewhere off the coast, close to Prince Rupert. They called for help, and everyone turned to Ben to help them. Never having sailed before, he drove his boat out to the sail boaters with a mate, climbed on board, and helped them sail the rest of the way to Alaska.

At fifty, Ben was now a successful salvage logger. He had a fleet of boats that included his twenty-eight-foot, welded aluminum crew boat, a sixteen-foot skiff for getting to shore, also welded aluminum, a massive tug boat I guessed to be fifty feet long, and a side winder for sorting logs in the booms. But the *pièce de résistance* was his ten-man floating camp, complete with fresh running water, generators, freezers, showers, and full galley with dining room.

Ben made his living at cruising the coastline looking for fresh fallen trees and logs that were either adrift or washed up on the beach. They had to be merchantable timber, in other words, logs that the mill back in Prince Rupert would accept for cutting into lumber, not driftwood. As I understood it, Ben also had a special license to cut down some standing timber along the coastline, with restrictions.

Once Ben found a merchantable tree, not standing, he would pin it, tie onto it, and haul it back to his floating log booms that were set up in Crow Lagoon, some thirty miles northeast of Prince

Rupert. These log booms were massive. Around September, before the weather became too inhospitable, Ben would haul the booms down the outside waters of the Pacific north coast to the mill in Prince Rupert with his more-than-capable tug. Hauling these massive booms through rough and treacherous seas and keeping them together was no small feat and not for the faint of heart. The log booms represented Ben's entire season of salvaging and were worth a considerable fortune, which he collected from the mill upon delivery.

Ben was no crude salvage logger, far from it. Every piece of equipment he owned shone like a new penny. All of his equipment was properly maintained and serviced by Ben himself, including his floating man camp.

There was not a spot of rust or a speck of dirt on anything Ben owned. He was, in every sense of the word, a complete professional. Also, fortunately for me, Ben was an excellent captain and fisherman.

It just so happened that Ben needed to do some work on one of his generators at the floating man camp up in Crow Lagoon around the same time I was to arrive in Prince Rupert. He asked me if I would mind going up with him to work on the generator and then go fishing from there. I was ecstatic and readily agreed; I'd never been to that part of the west coast before.

Crow Lagoon is an anomaly, like so many other special areas along the west coast. Crow Lagoon is actually considered one of Canada's top ten volcanoes and is still thought to be active, although no one knows for certain when the last eruption occurred. The entrance into the lagoon is almost completely shut off at low tide, making it a huge and very deep lake, but at high tide you can bring a boat in or a boom of logs out. It was inside Crow Lagoon that Ben had his log booms anchored to the shore and it was here too, that Ben had his man camp anchored to the steep walls of the volcano near the back of the lake, where a freshwater stream came into the lagoon.

The camp was palatial—at least it seemed so to me after having spent six weeks in the back of my camper travelling the Yukon. The floating man camp consisted of three separate trailers set up on a huge floating log boom, along with a shop and storage area. One trailer housed the bedrooms, six in all, with a separate bathroom area, with four shower stalls and toilets. The other two trailers were joined together and set up with a mud room, laundry room, freezers, an office, and then, on the end, the kitchen area and dining room.

Off the front of the man camp was a large deck area that had, among other things, like lounge chairs and picnic tables, a propane-fired beer keg which Ben used as a crab-pot boiler. He had crab traps hung off the front of the deck down about a hundred feet in the water, and each night when we came in from fishing he'd fire up the propane tiger-torch boiler, half full of sea water, pull the crab traps up, and throw the crabs directly into the boiling cauldron of sea water. I have never tasted crab so sweet and fresh. He also has prawn traps that hung off the front of the deck of the man camp, two hundred feet down. Without fail, they held multitudes of delicious, Pacific spotted prawns every time we hauled them up.

Ben was a very serious captain—everything ship-shape and all that. When I first got on board in Prince Rupert to make the two-hour run up to Crow Lagoon, he said there would be no drinking or smoking on board while we were fishing. I was a little disappointed to hear him say that, because while in Skagway I had found a store, one of those old-fashioned frontier-type stores you'd expect to see in a black-and-white cowboy movie, typical of the spirit of Skagway, that sold Everclear, 150-proof alcohol, and I had purchased two mickeys. You could light fires with that stuff. I'd also purchased, in the same store, two packs of what I called Clint Eastwood cigarillos, the thin, dark ones like Clint smoked in *The Good, the Bad and the Ugly*. I had specifically bought these in anticipation of our fishing trip.

The first morning of our trip, Ben took the smaller skiff and showed me around Crow Lagoon, which at the time I didn't fully appreciate but to which I have since longed to return. It was late August now, so his log booms were considerably full from his year of salvaging and hard work. Near the entrance of the lagoon, Ben put the skiff ashore and told me to follow him up into the bush. The trees along the shoreline were thick and dense, with lots of deadfall due to the constant winds, so it made walking through the forest slow and difficult. We hadn't gone far before we came to a tree that Ben wanted to show me, a Sitka spruce.

Compared to all the other trees around it, the Sitka was massive; the trunk of it was so large, four people holding hands would not have been able to encompass it. I looked up and saw that it towered above all the other trees surrounding it. Ben stood there smiling, seeing the astonished look on my face at the gigantic tree.

"Sure would like to have that one in my bundle," he said. "Just a little too far for me to reach from the boat though."

I knew my trees from being brought up in Revelstoke, and had heard of Sitka spruce, but I'd never seen one before. I saw the respect in Ben's eyes for the mighty tree, and I liked him right away.

We went back to the camp and switched boats to go fishing. Once we cleared the lagoon, Ben put out the rods and started trolling. Ben put on your standard green hootchies with flashers. After a while without any action, I politely and respectfully asked Ben if he'd mind if I fixed myself a small shot of Everclear and Coca-Cola, to celebrate our fishing trip, to which he snarfled but reluctantly agreed. So I poured a tiny amount of the Everclear, because that's all that's needed, in the bottom of a cup and filled it with Coke. Even then, it was barely drinkable. I showed Ben the bottle of moonshine I'd bought in Skagway, something we can't get in Canada, and asked him if he'd like a drink too. "Perhaps I'll have a small one," he allowed.

So I poured him a drink. After he took a sip, holding himself back from coughing, he said, "Whoa, that stuff's not bad, is it?"

Then I asked if he'd mind if I had a smoke—I was dying to light up one of my cigarillos. Again, he muttered under his breath but agreed, so I did.

"That smells pretty good," he said.

"Would you like one?" says I.

"Sure, let me try one," says Ben.

We both just stood there in the back of the boat, with the sun beaming down on us, without talking, trolling along through some of the most beautiful coastline in the world, with only the noise of the downrigger wires humming, smoking our cigarillos and drinking Everclear. The world was at peace and I was completely content. That was the start of a beautiful friendship.

The fishing was good but not overly spectacular. We were hoping to catch a big spring, a chinook salmon. Ben told me he had caught a forty five-pounder in this area before, but we weren't getting into any springs.

Off toward another island, around the corner from where we were, there was a coho opening and a fleet of trawlers were very apparent, fishing the area, so we worked our way over to them and managed to pick up a few coho.

We worked our way around the island where the coho fleet were fishing, then back down the channel to the camp, trolling back the opposite way we had started. Amazingly, we picked up two nice halibuts over some bumps on our way back, both about thirty pounds—"chickens," as Ben called them—perfect eating fish.

The tide was with us, and accessing Crow Lagoon was straightforward. As we hit the still waters of the lagoon, I noticed all the seals basking in the sun on top of the log booms. I pointed them out to Ben as we were coming in, and he said they were always there but recently there seemed to be more than usual.

That night I helped Ben cook up a scoff of coho salmon with some fresh crab and shrimp appetizers. I was in heaven. I told Ben I was going to take a dip off the end of the deck after supper, to freshen up. He warned me about how cold the water was, and, sure enough, it was as cold as he said. After I hit the freezing cold water, I scrambled for the safety and warmth of the deck again, pulling myself out as fast as possible, buck naked. Ben was having a good laugh, and much to my surprise had grabbed his camera and took a picture of me in my frozen nakedness. That picture ended up becoming somewhat of a longstanding joke between us, and without my knowledge showed up at my surprise fortieth birthday party in a 16×10 framed format for all to see.

The second day out, Ben said, "I'm going to take you to a special place today. We'll tie the skiff on at the back of the crew boat so we can use it to go to shore once we get there."

He explained that we were going to an area called the Khutzeymateen (pronounced "cootz ma teen"), and once there we would use the skiff to go ashore, to the Khutzeymateen River, and fly-fish for pink salmon which were running at that time of the year.

The Khutzeymateen Inlet and surrounding area is absolutely pristine, a natural coastal-marsh habitat for grizzly bears—the first area in Canada ever preserved for the exclusive protection of grizzlies.

When we first came into the inlet, we stayed offshore for a while, glassing with our binoculars for grizzlies. It didn't take long before we saw a sow grizzly with two cubs foraging along the shoreline in the thick, marshy reeds. I've always considered it a human privilege to see a grizzly bear. I have been fortunate in my lifetime to see many grizzly bears and appreciate ever single sighting. I love how my First Nation brothers refer to it as "*a gift from the Creator.*"

We anchored the crew boat out beyond low-tide waters and took the skiff ashore. I had purchased a new fly rod and reel specifically for some tidewater fishing, with a sinking line and floating tippet; this would be a perfect time to use it.

I followed Ben up the river, walking until we came to some deeper pools. The river was buzzing and splashing with pink salmon all the way up. The pools were black with schools of pinks. We call the pinks "humpies" because the male pink salmon develop a big hump on their backs, along with a hooked jaw, once they hit the fresh water of the rivers when they return to spawn.

It took me a few lost fish to get the hang of flossing for the pinks. The salmon aren't actually biting the hooks; they stop feeding once they go upriver from the ocean; they are in full spawn. The trick to catching them is what is referred to as flossing, whereby you drag the fishing line across their mouth until the hook lodges in the corner, then you set the hook and the fight is on! As challenging as that sounds, once you do it a few times successfully, you get the hang of it and can land a pink virtually with every cast. If by accident you end up snagging a fish—that is to say, hook it in any part of its body other than the mouth—it is not considered a catch. Of course, we released all of the fish we caught. Between Ben and I, we lost count of how many fish we caught, but I estimated it as well over twenty. Such a lot of fun! The tide was coming in now, and the fish were moving upstream, so we decided to call it a day; I called it a great day.

I have never been back to the Khutzeymateen Inlet, but I will. It is now a Canadian National Park, a Class A park dedicated to the protection of its natural environment. As I understand it, the local First Nations have stewardship of the park, so to go there now you must register with them at the floating park office. There is also a local tour boat that goes to the Khutzeymateen from Prince Rupert in the summer months.

We dragged the skiff back to camp, untied it, and secured it to the dock. Ben said we still had time to head back out and try for some springs—he knew of a couple of good kelp-bed spots out in the channel not too far from camp. So we did a quick turn-and-burn and headed back out again. The area and tides were perfectly set up for some good fishing. Chinook salmon like structure and kelp beds: if you can find both, it's reasonable that you will find fish there. Structure meaning underwater rocks and ledges.

I call Chinook salmon springs, as do many of my Canadian fishing friends. The Americans call chinook salmon kings. They call coho silvers, sockeye salmon reds, and pink salmon pinks. We all call chum salmon "dogs," the least preferred of the five. The chinooks are the largest of the salmon species and the preferred fish for angling. Chinook is my personal favourite to eat, but there are many that would argue for sockeye or coho.

It wasn't long after we started trolling that one of the rods had a hit. Ben, for whatever reason, grabbed the rod and started reeling it in. I suppose I thought up until that moment that I should have all the rods for some inherent reason, but once I realized Ben had a nice fish on, I was happy for him and the boat. It was a nice spring, about twenty pounds. It took him fifteen minutes to play it to the boat. I carefully netted it headfirst; high-fives and a happy dance were had by all. We reset the lines and continued trolling around in the same spot. We were in sixty feet of water, trolling along the shoreline, both lines at forty-five-foot depths. Bang, the other rod went off.

"There you go!" yelled Ben.

I grabbed the rod quickly, reeled in the slack, then set the hook. The reel screamed as the fish made a long and mighty run, straight toward the kelp beds. Ben quickly turned the boat out and away from shore, hoping the fish would turn away from the kelp. Luckily for me, the fish turned and ran back towards me and the boat. I cranked like crazy on the reel, hoping not to give the fish any slack line. No sooner had it made a run towards me then it turned and ran away from me again. I knew then we had it hooked well; it was just going to be a matter of time and keeping it out in the deeper water, away from anything that would break it off. After what seemed like a long time, we saw the back of the fish as it came to the surface and porpoised out of the water. It was a big, black-backed spring.

"That's a nice one!" Ben yelled out as it broke the surface.

The fish kept making runs and I kept reeling it back in until I finally I got it up to the side of the boat. Ben grabbed the oversized net and expertly reached out and netted it. Phew, it was a good size for sure. Ben got his scale out, and with a wry smile turned to me and proclaimed, "Thirty-six pounds! Nicely done!" The biggest salmon I had ever caught. What a day!

"Ready to head in?" Ben asked.

I grinned and nodded.

Ben, being the meticulous captain he is, had a stainless-steel fish-cleaning table on board that attached to the gunwale, at the rear of the boat so the guts and blood spilled overboard as he cleaned the two fish. I helped organize and clean the boat while we floated calmly adrift, making ready to head back. Ben filleted the fish expertly, four giant halves of gorgeous red salmon meat. He was almost done cleaning up, washing the last of the blood off the boat deck, when he spotted something off the stern.

"Look, blackfish!"

He pointed behind the boat. I looked back and, sure enough, could see a pod of orcas, spread out, slowly hunting, pack-like, toward us. I guess they were looking for the same thing we were, some fish of their own; obviously this was a good spot.

The pod disappeared, so we started the engine and went the short distance back to camp.

The tide was starting to ebb by now, but it still gave us plenty of water to clear the rocky-reefed bottom at the entrance to the lagoon. As we came in, I asked Ben if he would troll past the log boom so I could count the number of the seals. I did my best, but there were so many it was hard to be accurate.

"How many?" asked Ben as we cleared the end of the last boom.

"I count seventy-seven, could be more," I replied.

We docked, tied up the boat, and packed the fish into the freezer. The sun was still warm. I poured us a couple of Everclear and grabbed the last of my cigarillos, then we sat out on the deck of the camp; relaxed and happy from the day's adventure. Ben was a man of few words; we just sat there without saying much, looking out over the stillness of the lagoon. How could I thank him? I looked over at him and could see he was happy too, in the way that one is when they have no need for thanks. He knew!

The sun dropped behind the trees surrounding the lagoon, but the sky was still pink and red.

"Red sky at night, sailors' delight," Ben quipped. "We'll head back to Prince Rupert in the morning."

I looked over at him and nodded in acknowledgement.

Just then I noticed something off in the distance, near the mouth of the lagoon. It was the unmistakable sight of the dorsal fin of a big bull orca.

"Look!" I said to Ben as I pointed toward the entrance of the lagoon.

"Blackfish," said Ben.

Then we saw a few more spouts of spray, and dorsal fins in the calm waters of the lagoon from other blackfish. I noticed at the same time that more and more seals were jumping out of the water and onto the log booms. It looked to me like there were about a dozen or so blackfish in the lagoon now. I also noticed that the pod had divided into two groups. One group, consisting of the big bull I'd seen first and a couple of other blackfish, were positioned at the entrance of the lagoon; the other group, made up mostly of cows and calves, were swimming in a single line, coming up and submerging together like a well-trained platoon of soldiers on the march, heading straight for the log booms.

The tide was low now, and the entrance to the lagoon was almost completely inaccessible. The sentinel blackfish were perfectly positioned at the entrance of the lagoon, on guard.

All of a sudden, one of the bigger cows in the lead group, broke the surface in a terrifying leap that startled both Ben and me and thrust herself up onto one of the log booms. Seals went flying everywhere, like children on a trampoline. As they hit the water, we could see the waiting pod of blackfish grab them and tear them apart, one after another, over and over.

Soon other cows joined in on the log booms. As they landed, the ends of the logs would submerge under the weight of the clever predators, then they would wiggle their way back into the water, sliding back down off the logs.

Over and over the pod of blackfish launched themselves onto the log booms, until all the seals had either been flipped into the water or had jumped off in sheer terror.

What a spectacle we were witnessing. At one point, one of the smaller calves, learning to hunt, jumped up onto a log boom, repeating what it had seen its mother do, but the calf was too small to submerge the log boom, so it became temporarily stranded on the boom, trying with all its might to wiggle off, but unable. In an instant, another cow—I presumed, its mother—jumped onto the boom beside the calf, temporarily helping to submerge the end of the logs, making it easy for both to slide back into the waters of the lagoon.

It was truly amazing to see the coordination of the pod, working together to feast on the seals. There was nowhere for the seals to go except into the depths of the lagoon; only then to be prayed on one at a time by the hungry bulls that guarded the entrance.

We watched until it was too dark to see anymore. Of course, it was all we could talk about as we went inside and finally to bed. The next morning, we packed up and made ready to return to Prince Rupert.

The waters of the lagoon were still and peaceful, a complete contrast to the bloodbath of the evening before.

"Ben, troll by the log boom on the way out," I asked. "I want to see how many seals are left."

The pod of blackfish had long gone during the night, probably on the high tide. Ben trolled by the log booms on our way out. It wasn't hard to count the number of seals now. There were nine left.

CHAPTER 5

Camarones Café

I found myself sitting outside a little coffee shop, on a white plastic chair, at a white plastic table, the kind of table with a hole in the middle of it for an umbrella, drinking a strong, freshly brewed cup of coffee, in El Centro in downtown Sayulita.

It was about seven p.m., and El Centro was just coming alive with the hustle and bustle of merchants getting ready for the evening's activities.

I was doing some people-watching and relaxing after a strenuous day of reading and suntanning in the little Mexican seaside fishing village.

After a while my attention was drawn to a group of Mexican men having what looked like a tailgate party, grouped around the back of a Ford pickup, just down the street. They were laughing and joking with each other, drinking beer. They were just like a bunch of big kids, like I was when I was in my twenties, carefree.

The men's main focus was on something in the back of the pickup, and through their laughter I could hear something about *pescado*. I have a limited understanding of Spanish, but being a fisherman I knew that was the word for fish. So my curiosity got the better of me, and I walked over to the men and in my limited Spanish said hello and repeated the word *pescado*. "*Pescadores*? Are you fishermen?"

They were a fine bunch of guys, and they all started talking to me at once in Spanish, which I could not understand. Fortunately, among them was a younger man named Carlos who spoke very broken English. He explained that they had just come back from fishing and had been very successful. I asked him what they had caught, and he pointed to a gunny sack in the back of the truck. One of the other men, who I came to understand was the captain, or leader of the group, although I would hesitate to call him a captain, reached into the gunny sack and brought out one of the ugliest-looking bottom fish I had ever seen, with a tail on it like a ray, but it was obvious they were proud to have caught this fish. I smiled a big smile and said, "*Bueno*, very nice."

They could see my real enthusiasm. Chava, as the man was called (pronounced "chaw ba"), reached into the bag again and pulled out a couple more ugly-looking fish as well as a huge octopus. I could see how proud they were, so again I congratulated them. Chava offered me a beer. I still had half a cup of coffee back on the plastic table, but I did not want to insult them by refusing their kind offer, so I accepted their beer with great appreciation.

A very awkward conversation ensued, a mix of broken Spanish and broken English, as we drank our beer and joked around. Carlos did the interpreting for the group. Chava asked me if I liked fishing; I told him I had done a lot of fishing in Canada, including ocean fishing for salmon and halibut.

I was introduced to the other two amigos, Pedro and Hernando. Hernando, Pedro, and Chava were all three-hundred-pounders, big Mexican men, but they laughed and made fun of each other like little boys. I liked them all right away, especially Chava.

After I finished my beer, I said goodbye and walked back to my table. Next to the coffee shop was a little Mexican grocery store. I walked right past my now cold cup of coffee and went into the store and bought a twelve-pack of the kind of beer my new amigos were drinking, Pacifico in dark-green bottles. That case of beer turned out to be one of the best investments I ever made. I walked back to the men and gave them the case of beer, as a gesture of congratulations, from one fisherman to another, for their hard-earned catch. They accepted my offering and insisted I stay and have another beer with them. That was the beginning of a beautiful friendship, one that will endure for the rest of my life.

Chava asked me if I wanted to go fishing. I told him I'd be honoured.

"When?" I asked. He spoke in indiscernible Spanish but through Carlos I learned that they were going fishing the next day, *mañana*!

I said, "Sure, I'd love to," finished up my beer, and asked again, just to confirm, "*Mañana, si*? Where do I meet you?"

Through Carlos I was told to meet at Chava's home and given the necessary directions.

I rushed home all excited about going fishing with some real Mexican fishermen. I had paid for fishing excursions in Mexico before with limited success. I had enough of an understanding to know which months of the season were best for fishing, and what that meant in Mexico. It meant, the months when the tuna were in, *atun,* big tuna, some in the vicinity of weighing up to two hundred kilos.

As soon as I got home, I prepared my camo pack as I always do. I made a lunch, which I put in the fridge until morning, made sure I had water, a cap, sunglasses, sunscreen—the basics for a hot day out on the ocean. That night I went to sleep dreaming of big tunas!

I awoke with great anticipation, around six a.m., ate a quick breakfast, prepared the last of my camo pack and made my way over to Chava's on foot as instructed. There weren't many people out on the streets that early; finding Chava's home wasn't difficult. When I arrived shortly before seven a.m., there wasn't a soul in sight.

Chava's house was in a busy part of Sayulita. He and his wife had fashioned it into a restaurant, with tables, chairs, and a wood-fired oven on the front patio, facing the street. I was pretty sure I was at the right house, so discreetly—or, more to the point, respectfully—I walked across the patio and knocked on the front door, a big wooden, iron-clad door. There was no answer, so I thought better about knocking any more and decided to wait until someone arose and came out.

I sat at one of the tables and waited, and waited, and waited. At eight o'clock I knocked on the door again, but there was still no answer, so I waited some more.

Around nine o'clock, Pedro came out, said good morning to me in Spanish in a still sleepy, groggy voice, and plunked himself down at a different table from the one I was sitting at. Shortly after that, Chava came out, and in a hungover voice said, "*Hola, amigo. Buenos dias. Pescador, si?*"

He waved me over to the table where Pedro was sitting, near the cooking fireplace at the back, and proceeded to sit there himself. I joined the two men, shortly thereafter Carlos and Hernando joined us, and we all sat around talking and joking like we had the previous night. I had no idea what the plan was or why no one was in a hurry to go fishing, but I have learned one thing about travelling in a foreign country: watch what others do and do the same.

Then a lady appeared through the front door with a plate of tortillas and rice and beans. Chava introduced her to me as his wife. I realized they were all about to eat breakfast. Chava's wife continued to bring out breakfast, which included scrambled eggs, chorizo sausage, and fruit. It wasn't long thereafter that the first beers were opened. I had declined breakfast because I had already eaten and did not want to risk sea-sickness out in the boat, but I felt that if I was to fish with the boys I had better drink with them too, so by ten o'clock we were into the *cerveza* again.

After breakfast Chava wandered off while the rest of us sat around talking and drinking beer. After a bit, Chava came back in a white Dodge pickup, either his own that had been parked somewhere else, I thought, or one borrowed from a friend. Everyone jumped in the back, and I was about to join them, but Chava insisted I ride up front with him. I always chuckle to myself, incidentally, about Dodges in Mexico. The Mexicans pronounce the word Dodge as "doe-hay," the *g* becoming a guttural *h*. The first time I heard them saying it, I had no idea what they meant!

Chava drove the truck uptown, back to El Centro, and parked it in almost the exact spot where we were tailgate-partying the evening before. Then they all jumped out and went into the little Mexican grocery store from which I had purchased the case of beer. I figured they were going to get some food for lunch for out in the boat—and some more beer, of course.

Carlos came out first with a fifty-pound sack of onions and threw it in the back of the truck. Then Hernando came out with a fifty-pound sack of roma tomatoes. Then Pedro, with a sack of limes, a big sack. Then Chava with two cases of Pacifico. Then they all went back into the store for more provisions. They filled the back of the pick-up with sacks of vegetables and fruit, a huge wholesale sack of extra-large Mexican taco chips, and beer, lots of beer, all the same kind, twenty-four packs of Pacifico in green bottles, in big heavy boxes.

I thought to myself, *Holy cow, these guys can eat and drink! No wonder they're all three-hundred-pounders.*

I offered Chava five hundred pesos, which is about fifty dollars, but he refused to take it. I insisted, saying it was for the *cerveza* and *gasolina*. He reluctantly accepted it, thanking me, then everyone jumped back into the truck, and down to the beach we drove. It was now well after eleven a.m.

When we got to the boat—a twenty-six-foot *ponga* equipped, like all the other *pongas* on the beach, with a ninety-horsepower, two-stroke Yamaha outboard motor, I had a look inside. It was an open boat, without even a sunscreen canopy. There were no fishing rods in it. There was no bait. There was no net or gaff, not even a cooler to put the fish in. There wasn't a single life jacket. I was puzzled, wondering if this was really the boat we were going fishing in, but all the boys started packing everything out of the truck and into the *ponga*. So, following my rule of travel, I threw my pack into the boat and went back to help carry all the beer and groceries down to the boat.

Once all the store-bought provisions were in the boat, Chava summoned everyone he could find on the beach, who all appeared to be his *primas* (cousins), or at least his friends, to help us push the boat into the surf.

Chava said something loudly to Carlos, who interpreted instructions to me that once we hit the water, I was to jump in the boat first.

We bent our backs and pushed the heavy *ponga* off the beach, and, once the first wave hit the boat and it began to ride up on the surf, Chava yelled at me to jump in, which I did eagerly. One by one, the other men all jumped in as we hit the second and third standing waves, then lastly Chava jumped in, and with one pull of the start cord fired up the motor with a loud roar just as a fourth wave soaked the boat and everyone in it, and steered the *ponga* out into the ocean. Once we got through the standing waves, the ocean became calm, and it was easy boating from then on.

Chava headed the boat straight out from the shore at full throttle. The bench seats, which reached from one side of the boat to the other, lifted up to reveal storage spaces; however, none of the sacks of vegetables or the other provisions, which were lying strewn across the bottom of the boat, was placed in the seat storage; this space was reserved exclusively for the beer. The boys began dumping bags of ice in the seat lockers, opened up a couple of twenty-four cases of beer, and filled the seat lockers with as many bottles as they would hold. Then we all began to drink beer again as Chava headed us out to sea while the shoreline began to fade out of sight.

In my pack, among many other necessary items, I always carry a Bic lighter, in particular a white Bic lighter. Little did I realize that this lighter would become the tool of choice for all. Thanks to it, I was initiated into the fraternity of Mexican fishermen with a lifetime membership—well, on this boat anyway. The Pacifico bottles had hard caps on them, not the screw-off type. I brought out my Bic lighter to open my bottle, then offered it around so everyone else could do the same, which they all did. Moreover, all of them smoked, except for Chava and me, so they used the lighter for their cigarettes as well. My Bic lighter was constantly in demand throughout the day, and I relished being asked constantly for the use of it. It made me feel like I was contributing to the team.

Onwards and onwards we went until, looking back, I could no longer see the coastline, only the horizon. Then without warning, Chava slowed the boat down to a stop and shut the motor off. The seas were calm, and the day was beginning to become hot. They all yelled something to me in Mexican. I looked at Carlos again for interpretation.

"You want to have a shower?" he asked. I shrugged my shoulders and said, "*No comprende.*"

He said, "You jump in the ocean."

I'm not sure what the look on my face was at that moment, but I saw all of them looking at me and silently waiting for me to do something. I glanced back to the shoreline which had now disappeared, and thought to myself, *there is no way I am jumping in the ocean way out here, and you guys take off on me and leave me to drown.*

Collecting myself, I said to Carlos, "*No, gracias,*" as calmly as I could.

So Carlos ripped off his shirt, stood up on the gunwale of the *ponga*, and dove in, then one after another, they all did the same. Realizing they weren't out here to do me in, but merely to wash off the previous day's sweat and grime, I peeled off my T-shirt and followed suit.

We were all splashing around in the water, laughing at each other, miles out in the ocean, with none of us being particularly good swimmers; Pedro, being the worst, was barely able to keep his head above the water.

Once we'd had enough, we all swam back to the *ponga* and with great effort pulled ourselves up into the boat, which was somewhat challenging to do because of the high gunwales on the *ponga*. We all managed to pull ourselves up and into the *ponga*, all of us, that is, except for Pedro. He was spent from swimming around in the ocean and could only manage to cling onto the side of the boat for dear life. We all grabbed for his hand or an arm and tried pulling him into the boat, but he was heavy, each time we lost our grip, and he fell back into the ocean and went under. Each time he came up, he could hardly breathe and was barely able to reach up and grab the gunwale of the boat again. There was no way he could pull himself up into the boat, and somehow we could not lift or pull him up into it either. The worst of it was, every time Pedro fell back into the water, and disappeared, he stayed under longer and longer. We all collectively held our breath, waiting for him to come up each time.

Finally Chava came up with an idea. We reached down and grabbed Pedro's flailing arms and pulled him up high enough for him to hold onto the side of the *ponga* again. Then we all leaned over the same side of the boat, which made the normally very stable *ponga* tilt over in the water, downward towards Pedro. This manoeuvre enabled us to grab more of a piece of him, and with much difficulty we managed to roll him into the *ponga*, where he flopped onto the bottom of the boat like a giant tuna we had just landed. He lay there gasping for air!

We were just killing ourselves laughing, I had tears running down my face, and after a few minutes, when Pedro could breathe again, he joined in the laughter. Pedro wore a big straw hat that made him look even bigger than he was. Well, that called for a beer, and a smoke, and a Bic lighter, and more beer.

Chava then fired up the motor, turned the boat ninety degrees, and at full throttle headed off again. I thought to myself, *Finally, we are going fishing,* but how and with what gear I was not sure.

We sped along for about half an hour, until off in the distance, I could make out the shape of a ship toward which we seemed to be heading. As we got closer and closer, I could see the ship was at anchor, out in the middle of nowhere. It was a scow, a derelict kind of pirate-looking ship, with yardarms, masts, and riggings unkempt, hanging off on both sides. Perched on every available yardarm or mast were one kind of sea bird or another: seagulls big and small, cormorants, albatrosses, frigate birds, and other species I had never seen before. Hundreds of birds, all sitting and pooping all over the boat, further contributing to the derelict-pirate motif.

Chava came up to the ship and circled it, yelling at anyone on board who might hear, but no one came out on deck or answered back. It was deserted, like a ghost ship, I thought to myself. Finally, we came up to the rear of the ship, where in true pirate style we tied the *ponga* up to the woven rope ladder on its stern.

Once Carlos and I, being the nimblest of our crew, climbed up the rope ladder and were up on the deck of the derelict scow, the rest of my amigos started handing us up the supplies, including the many cases of beer. Carlos and I stacked them in a pile neatly on the deck. That done, Chava, Pedro, and Hernando climbed up the rope ladder and came on board as well.

Chava waved me over, I followed him to the stack of cases of beer. He motioned for me to grab a case. I followed him over to a vat on the main deck, where we had stacked the supplies, and where now each of the others had taken a seat and opened yet another beer. Chava, who did not speak a word of English and was never going to try to learn, lifted up the lid of the vat, which looked for all intents and purposes, like a giant cooler, and showed me the contents. Inside the vat, or cooler, was a dark green, salt-water-like liquid that was extremely cold. I'd never seen anything like it. Hanging above the vat was a mesh bag. He took the beers from me, removed them from the box, placed them in the mess bag, and lowered the bag into the green brine, which was so frigid that if you put your hand in it for thirty seconds, you would suffer irreparable frostbite. I remember thinking to myself it was the opposite of a microwave oven—it made the beer cold instantly.

Once the beer had been taken care of, we went back to the others and joined them sitting on the deck. Well not exactly on the deck; we sat on tiny stools. They were of the simplest design, like most things in Mexico, but very effective. The little stools were made of a wooden board on top to sit on, about twelve inches in width, and underneath at each end, were two legs consisting of pieces of wood that were rounded on the bottom. The height of each stool was only six inches. The ship was constantly swaying back and forth, rolling from side to side, even in the relatively calm sea, so the stools with their round-bottomed legs allowed us to roll along with the motion of the vessel—a perfect design.

We had been on the ship now for about half an hour or so, and yet still none of its crew appeared to be around. I felt uncomfortable, like we were drinking on someone's ship without their consent or knowledge. This, however, didn't seem to phase my Mexican amigos as they sat there on the deck drinking beer after beer, smoking, laughing, and enjoying themselves. All the while, the birds above were pooping everywhere, including on us. At first I was taken aback by the droppings landing on my head, but it was happening to everyone else too and they didn't seem to mind, so there I sat, rolling back and forth on my little stool, with sea birds pooping on me. Of course, there was bird poop all over the deck as well, making it precariously slippery to walk on. Best just to sit, roll, and drink, I thought.

Finally, a man appeared from below deck, and greetings were given all around. Then the man started carrying the supplies back down from whence he'd come. After he had carried all the supplies away he came over to me and spoke to me in Spanish. I looked at Carlos, who said, "You go with this man."

The man, perhaps the captain as I thought, took me down some metal ladders and stairs to the hold in the bottom of the ship. He reached down and opened up a big trap door and shone a light inside. To my utter amazement, the entire hold in the bottom of the ship was filled with *camarones café*, brown shrimp. This was a shrimp boat, and apparently the crew worked at night and slept during the day. The man I had taken for the captain turned out to be the cook. He explained that *camarones café* are the most desirable species of shrimp. They are not as large as *camarones azul*, the blue shrimp, but they taste the same or better, and as he demonstrated, their shape makes them easy to peel, hold onto, and eat. We climbed down another derelict ladder into the hold until we came to the top of the shrimp. The cook handed me a giant woven basket to hold while he scooped up a massive pile of shrimp with a wire dip-net, like those used to clean swimming pools, then dumped the shrimp into the basket as I held it for him. Once he felt we had enough shrimp in the basket, we climbed back up out of the hold, he closed the trap door, and I followed him back up to the main deck. I went back to my stool and my ice-cold beer while he disappeared back below.

After a bit, the cook came up on deck again with a large bowl of food and a big bag of taco chips, the ones we had brought him from the store in Sayulita. He handed me the bowl first, treating me like an honoured guest, as did my crew mates. I was incredibly moved, not just by the moment but by the entire day, and the fact that my amigos held me in such high regard.

To my surprise, the bowl was not full of shrimp, it was filled with a cabbage salad made with fresh cooked tuna and mayonnaise. I was starving, having not eaten all day while drinking copious amounts of Pacifico. I dipped a taco chip into the salad, scooped out a generous portion, and passed the bowl around. It was the most delicious salad I had ever tasted, and there was more than enough for everyone. Over and over, as my turn came around, I scooped up more of the salad with the hard taco shells. Meanwhile the birds continued pooping all over us and into the salad; it was all thoroughly delicious and satisfying.

When most of the salad had been consumed, Chava handed me the bowl and motioned for me to finish it. I didn't see what there was to finish other than the lemony, tuna-infused mayonnaise brine on the bottom of the bowl, but Chava insisted I finish it, indicating that it was very good for me—healthy, I thought he meant. I could see that this was a delicacy they all prized and yet, in their gracious manner, were leaving for me, so I took a last taco chip and dribbled some of the juice onto it, making sure there was enough left for Chava, then gave him the last of it to finish.

Back to our stools, more bird poop, more beers, still more taken to and from the cooling brine, more cigarettes for the smokers, and more rocking back and forth and forth and back. The motion was continual. I suppose at some point you get used to it, but you would never forget it, your every movement being balanced with the motion of the ship and the rolling of the ocean. Shortly after we had finished the salad, the cook appeared again, this time with a huge bowl of cooked shrimp, and as seemed customary now he offered me the bowl first. I looked at the others, not sure what to do, they motioned me to grab a big handful and put it on the deck in front of me, so I did. I started to peel my first shrimp, but Chava waved at me to eat the whole thing, shell and all, so I did. The sensation of sitting on that stool, under the hot Mexican sky in the middle of nowhere, rocking back and forth with birds pooping on me while I ate *camarones café* is one I shall never forget. After a while I noticed Chava was peeling his shrimp and so was everyone else, so I peeled mine as well. The shrimp were so fresh the peels came off like wrappers on a candy. We ate and ate and ate. Piles of discarded shrimp shells accumulated in front of each of us as we gorged on. It was a splendiferous moment!

There were plenty of shrimp to go around, but eventually we were all so full we couldn't eat another bite. I wondered what was going to happen next. I felt like we had overstayed our welcome, but the boys kept drinking. Then the real captain appeared and greeted Chava. I could tell the captain was a no-nonsense kind of man, unlike my crew mates. After greetings were exchanged, the captain went about the deck doing this and that, then started up the ship's engines. Time to leave, I thought, but no, my amigos were still sitting and drinking, so I sat too. A second ship's crewman came on deck and helped the captain haul in the anchor. Our time was clearly up, so we slowly made our way to the back of the ship and climbed down the rope ladder back into the *ponga*. Between the rocking and the beer, this was no easy feat; I was peely wogged!

We untied from the ship and slowly drove away from the shrimp boat, waving to the captain and crew, who were hurriedly getting their ship underway. Chava turned up the throttle to full blast and headed back towards land. It was nice to be back in the relatively stable confines of the *ponga*. But we hadn't gone more than ten minutes from the ship when all of a sudden the motor started sputtering, then stalled completely. We were now adrift somewhere in the Pacific ocean. I looked at Chava, thinking he had done this deliberately. He guessed what I was thinking, so he yelled over to me, "*No gasolina, no gasolina.*"

What? *Great, I thought, just what we need now that it's getting dark.* I looked around the boat, but there was no more gas, no spare tank, nothing. *Now what do we do?* I wondered. Without breaking stride, the mates all whipped out their flip-top cell phones and started calling their friends back in Sayulita.

"*No gasolina, no gasolina, cerveza, mas cerveza,*" they said over their phones. Then everyone sat down, opened another beer, and started laughing and drinking again, so what could I do but laugh and drink along with them?

They were such happy people, the likes of which one doesn't often meet. I wasn't sure what the plan was, but in no time we could hear the sound of a distant motor, and up beside us came two other *pongas* with Mexican fishermen in them, offering us gas in exchange for beer.

The one fisherman who actually brought gas for us, and who was obviously one of Chava's *primos*, was called Chingaling. (Later on I came to know him well, but that is a story for another time.) He was a great guy, called me *gringo* right off the bat, which isn't exactly a compliment from a Mexican. Chingaling siphoned the gas he had by sucking up the fuel from one of his tanks and running it into our tank through a small rubber hose, obviously something that was a common occurrence in Mexico. We handed Chingaling a bunch of beer, started up the motor, and raced for home.

When finally we could see the tiny beach of Sayulita, Carlos said to me, "We make big crash on beach, you hold seat, put feet up like this," indicating that I should place my feet firmly against the seat in front of me.

"Then you jump out and pull boat up quick, quick."

Chava came racing in at full throttle, riding the crest of a wave, on our final approach, and just as Carlos had warned me, we hit the beach with a thud and came to a sudden stop. Out jumped everyone, including me, and grabbed the *ponga* so that when the next wave hit us from behind we were able to pull the boat further up onto the beach, and likewise with subsequent waves until we couldn't pull it up any further. Then Chava went up and got the Dodge, tied a rope from the truck to the *ponga*, and pulled the boat up all the way to the top of the beach, past the high tide mark.

I was whipped. I thanked my amigos and said my farewells. I staggered home like a drunken sailor, not from the beer so much as from the constant motion of the sea. When I got back to my room I flopped onto the bed and crashed, thankful for not having thrown up from sea-sickness.

The next morning I went over to visit a friend in Sayulita. He lives there; he is a Canadian *gringo* like me. I told him my story of the previous day's events. He listened and said afterwards that what I had experienced was an incredible honour. Even the local Mexicans do not get to go out and re-supply the shrimp boats; only a select few elders of the community do. He said that if I knew Chava, whose brother just so happened to be the acting mayor of Sayulita at the time, and if Chava had invited me out with him to the shrimp boat, then I was considered to be one of his family. An honour I cherish to this day.

Muchas gracias, my amigo Chava.

CHAPTER 6

End-of-Season Whitetail

The date was November 30th. I remember it well because it was the last day of the hunting season. My business partner and I had scheduled a meeting for that day, but my mind really wasn't on a meeting, it was on hunting.

I called my partner the night before to see if we could cancel our meeting for the next day.

He said, "Sure, no problem, but why?"

I told him that November 30th was the last day to hunt for whitetails. To my surprise, he suggested we go hunting together. He said he would drive, I could shoot. It also gave us the day together, driving around, to go over our business plans.

I had never hunted with him before, we didn't have that kind of a relationship; ours was strictly a business-partner relationship. Our business meetings were usually informal, more pep rally than official company business and usually involving lunch and many cups of coffee.

My partner was not really a hunter, although he had related a story to me about a hunting trip he had been on in the Kootenays some years back, before he was married and settled down, where he had a face-to-face encounter with two juvenile grizzly bears. Fortunately, everything worked out for him. The grizzlies ran away in the opposite direction!

True to his word, my partner picked me up bright and early the morning of November 30th, so we could make the six o'clock ferry at Shelter Bay. My plan was to head over to the Trout Lake area to look for whitetails. I knew at this time of the year we would encounter snow on our travels, but that would not deter two boys from Revelstoke. Snow is a given when you live and play in Revelstoke, one of the highest-snowfall areas in Canada.

Sure enough, once we got across the ferry and turned north to Trout Lake, we were into winter-land. A heavy layer of snow had fallen, but the roads were drivable, and I knew the fresh snow would help us see deer tracks easier. So, all in all, it was setting itself up for a good day of hunting—at the very least a good day for a drive and for us to get out of the office together.

We drove up to Trout Lake, about an hour, without seeing anything. I knew the area well; I had worked for the department of highways at Trout Lake for the previous four winters, ploughing snow. I was familiar with all the roads. Once we got to Trout Lake, we turned down the Gerrard Road and followed it until we reached the end of the lake. My plan was to turn around after the Gerrard Bridge and come back.

We drove down the thirty miles of winding Gerrard Road. Officially, it is Highway 33, but few people ever drive the road other than loggers and highway workers. Highway 33 eventually comes out at Kaslo, and from Kaslo you can get to Nelson, though it's certainly not the preferred route to Kaslo. The best route to Kaslo is through Nakusp and New Denver, which is an all-paved road unlike the gravel road of Highway 33.

There were a few deer tracks here and there where I had expected to see tracks, but no sign of any deer. Trout Lake is not a natural area for whitetails, so I wasn't too optimistic about seeing them there. The mule-deer season had already closed so I was restricted to hunting whitetail bucks only.

We came to the famous Gerrard Bridge, which crosses over the tail end of Trout Lake and then flows into the Lardeau River. Here at the head of the Lardeau River, which joins Trout Lake to Kootenay Lake, is the spawning grounds of the world-famous Gerrard rainbow trout, who make their way up from Kootenay Lake every spring to spawn in the ideal breeding grounds under the bridge at Gerrard. What makes these spawning beds so ideal for rearing giant rainbows is the cold, clear water that runs out of Trout Lake over clean gravel beds. It is a unique and special ecosystem, conducive for rearing the giant rainbows, an area I have visited without fail for the past forty years in the spring, to see the miracle of the rainbows as they breed, spawn, and then return back down the Lardeau River to Kootenay Lake. Some of the giant Gerrard rainbow trout reach in excess of thirty pounds, and have been known to return to spawn three times, over a twelve-to-fourteen-year life span.

We stopped on the bridge to look into the clear water below to see if there were any signs of life in the river. Sure enough, we could make out a large school of suckers spawning in a gravel bed off to our right. We could see the constant flash of the white markings on the fish as they built their reds in the gravel beds, laid their eggs, and fertilized them. I hadn't realized there was such a big population of suckers in the lake, but I suppose they are as natural to the lake as trees are to the forest.

We went over the bridge to turn around and head back, but as we crossed it, I noticed a new logging road on the right-hand side, just past the bridge, going back up the west side of the lake, one I had never been on before. I suggested we drive up it as far as we could.

There looked like some recent sign of human activity on the road. The snow had been packed down by other vehicles and was easy enough for us to drive on. We drove up the new logging road for a mile or so and came across elk tracks, obviously a crossing where the elk came down to the lake to drink.

We were carefully working our way up the new road, the sun directly in our eyes, when I spotted something.

"Stop, stop, look there," I pointed out.

Right in front of us on the road, coming over the top of the hill we were heading up, was a rabbit—or, more precisely, a Columbia snowshoe hare. (I call ducks ducks and rabbits rabbits!) It was hopping down the middle of the road in the previously made tire-track ruts in the snow, coming toward us as if we weren't even there.

I told my partner to stop the truck and shut if off. The rabbit came closer and closer until it was right to the front of the truck, then we couldn't see it anymore. I thought it was going to hop under the truck and keep on going back down the road we had just come up, but as I watched for it in my rear-view mirror to see it come out from under the back of the truck, I noticed the rabbit jump off the road on my side, and hop over to the only growth of vegetation there was in that area, which was a small clump of bunch-grass on the side of the road.

The rabbit stopped in the middle of it and sat motionless, hoping, I thought, that we couldn't see it.

The snowshoe hare was beautiful, even magnificent, you could say, as any prime animal of any species is. He was in his complete change of clothes by now, completely white except for a little black tip on his tail, which I noticed right away he carefully tucked under his rear so as it was not showing. He was so close to my side of the truck, I could see his eyes sparkle and his little pink nose twitch as he breathed—and he was breathing hard and fast.

I said to my partner that I could see him right there, just off to the side of the road. We could see the rabbit through the passenger window, hiding in plain sight in the clump of grass, in the snow, beside the road.

"What's up with that?" he asked. "What's he doing?"

"I don't know, but let's just sit here for a bit and watch him," I answered.

Somehow, I've been blessed me with a gift of seeing wildlife, and encountering special and unique wildlife events, for which I am truly grateful. This was one of those moments.

We sat in the truck for a while, perhaps five minutes, watching the rabbit sitting there like a statue in the grass pretending he was invisible. The sun was brilliant, I could see his fur glisten like thousands of tiny diamonds sparkling from the snow on his coat.

Then off to my periphery I noticed something move, which drew my attention back to the road.

"Look, look down the road," I said.

And then it became clear the scenario in which we had found ourselves. Hopping down the road, coming toward us in exactly the same tracks as the rabbit, was a full-grown, mature pine marten, obviously hunting and chasing after the rabbit. The marten was also a prime animal, with a full coat of guard hair shining in the sun. With his blonde highlights shading into red and dark brown markings, it is no wonder the marten is a favourite of trappers for its beautiful pelt. I have seen many martens in the wild—"bush kitties" as my one trapper friend calls them. A smaller cousin to the wolverine and fisher, martens are carnivores that primarily eat squirrels and rabbits, but have the ferocity of their larger cousins and are not to be trifled with.

Martens will eat anything they can catch, including birds and aquatic animals. I remember a good friend relating a story about a marten to me, when he was up in the bush logging. He had stopped for lunch and saw a marten sneaking through the trees to try and catch a crow that had landed in the tree next to him, the crow hoping for some lunch scraps. The marten snuck along a branch of the tree, above the crow and made a dive for it, but just as the marten grabbed at the crow's wing feathers, the crow flew away, narrowly escaping.

Our marten was after a particular snowshoe hare this day and he was not going to be deterred by some truck in the middle of the road. The marten, hopping in the way they do, approached us just as the rabbit had. We sat in the truck, motionless and speechless, not sure who to root for, the rabbit to get away or the marten to kill.

Truth be known I was rooting for the marten just as I do for all predators, like I would for a falcon over a bird, even though I love birds.

The marten, after getting to the front of the truck, hopped up the snow bank on the driver's side, opposite the rabbit, went around the truck and then back down onto the road behind the truck, in our tire tracks, and carried on down the road in the way we had come. I could see the marten behind us in the rear-view mirror on the passenger's side hopping down the road, looking for the rabbit.

I could also see the heightened awareness of the rabbit, who was morbidly still but with his nose twitching as fast as his heart was beating. I'll never forget his nose, pointed up in the air, nostrils flaring in and out with each breath, eyes wide open. The marten hopped out of sight.

"Wow, that was spectacular," I said. "Something you only see on National Geographic. I can't believe he didn't see the rabbit."

The rabbit hadn't moved. "I wonder what the rabbit is going to do now. Let's wait a bit, don't go anywhere."

I had barely gotten the words out of my mouth when I caught glimpse of another movement. It was the marten. I saw him in the rear-view again, coming back toward us. Our marten must have realized that he'd lost the scent and turned around.

Again in suspense, we watched in the mirror as the marten returned back up the road to the truck. This time, he came around on my side, hopping through the deep roadside snow only a few feet from the rabbit, who sat motionless in the clump of bunchgrass. Then back onto the road. Then the marten did a pattern of reconnaissance circles around the truck, up on the driver's side, back up the road whence it had first come, then back down, circling to within a few feet of the rabbit. The rabbit, as white as the snow, never once moved.

After what seemed an eternity but was more likely about fifteen minutes, the marten, unable to pick up the scent or sight of the rabbit, turned and left up the road in the same direction from which both animals had come. It was one of those moments you feel privileged to be a witness to, the purpose for Being.

"We should turn around and let nature take its course," I said.

I knew that sooner or later that marten was going to get that rabbit, but for now the rabbit was alive and that's the way we left it. We turned around and headed back to the Gerrard Bridge.

We drove back to Trout Lake in silence. There was no need for superlatives, we had both seen the same thing at the same time, unlike sometimes when you split up to go hunting and one hunter comes back and tells of an incredible sight they alone had seen.

A little further along, I said I knew of a road closer to the ferry that we should drive down. Even though it got dark early these days in late November, we still had enough time to make it in and out before it was too dark to shoot.

We were about an hour away yet, both from the ferry and the darkness.

We got off the gravel road of the Gerrard highway and back onto the paved highway from Trout Lake to the ferry. At the Hill Creek turnoff, we turned right and drove down the road, past the Kokanee spawning channels, to the end of the road, which in essence was a landing at the bottom of a logging block, just up from Arrow Lake. We could see the lake off to our right as we drove in. There were no tracks or sign of deer anywhere!

When we got to the end of the logging road, I asked my partner to turn around on the landing and shut the truck off; I had to relieve myself. I got out on the passenger side and started to take a leak. As I was doing so, I looked up into the logging block, which was bare from leaves by this time of the year, and standing halfway up the cut block, in among a bunch of willows, I could see the figure of a deer. Right away my heart jumped; I could tell it was a buck even though I couldn't fully make out its antlers. My years of experience told me it was, to be standing there alone like that!

"There's a deer right there," I whispered. "Look, can you see it? Don't move. Where are my binoculars? It's a buck, I can see it, it's a big buck, and it's a whitetail! I'm loading my gun. Don't move."

I always carry two shells in my hand or have two shells within easy reach when I'm hunting. My 30-06 doesn't have a clip, so I have to load the two shells through the top, into the chamber. I have done this many times and can load my gun as fast as anyone can with a clip. And I always load two shells, just in case.

The whitetail buck was standing about 250 yards away, uphill and quartered away from me, with his head turned looking over his shoulders, back downhill at me. I rested my gun on the open passenger-door window and found him in my sights. Up until then, I had considered myself a good marksman.

At 250 yards, the deer in my sights was tiny, for lack of a better way of describing it. All I was presented with for a shot was his rump, the back of his neck, and his head. Basically, I could see his ass, neck, and head as he peered over his shoulder back down at me. He stood perfectly still, probably thinking I couldn't see him through the willows.

I aimed over his back for the base of his neck, thinking that if I came up short, I would at least hit his body somewhere, probably right in the ass in a worst-case scenario. How right I was going to be—worst-case scenario, that is!

I have a friend who has a huge whitetail mounted in his house. I had the honour of seeing it one day. It is a two-hundred-point typical. I asked him where and how he had shot it. He said he had been walking in his secret spot and just came upon it. The deer saw him, he saw it, the deer turned to run away, and he said, "I just shot it in the ass." And that was his story.

I was not thinking of his deer as I aimed. I was thinking of making a perfect shot. I squeezed the trigger and fired. Nothing! The deer just stood there. I loaded my second shell into the chamber, carefully aimed in the same spot, and fired again. This time the deer turned broad side and looked back at me. He could obviously see the truck.

"I'm out of bullets, I'm out of bullets," I said in a slightly panicked voice.

"What, we came all this way out here and you only have two bullets?" my partner said in disbelief.

"No, no, I have a box of bullets in my pack. Where's my pack?" I ripped open my backpack and grabbed the box. This time I grabbed four shells, the most I can easily load into my 30-06. I reloaded all four bullets, reset my gun through the open door of the truck, and took aim at the deer. Boom, I fired off another round in haste. Now the deer starting walking downhill, toward us.

Boom, boom, boom, I fired off the other three bullets I had chambered.

"More bullets, I need more bullets," I said in a more heightened, panicked voice.

"Stop! Look!" my partner yelled. "He's walking down to the road. He's going to come out right there on the road in front of us, once he crosses through that gully. Stop shooting."

I realized my partner was right. The deer was walking down the hill towards us, and was going to cross the road right in front of us, about a hundred yards away.

"Okay, okay, I'll shoot him right there on the road as he crosses." I could not believe my eyes, or luck.

I reloaded four more bullets. The overgrown road was straight for as far as I could see in front of the truck. I re-positioned myself through the truck door window and aimed down the road, waiting for the buck to come up through the ditch, on the high side of the road, and cross in front of me. I waited and waited. Sure enough, the buck broke through the brush on the high side and stepped down onto the road in clear sight, broad side. I calmed my frayed nerves and waited until he was fully in the middle of the road, took aim just behind his front leg, and fired. Nothing!

The buck leapt over the road and down the bank on the low side. I quickly chambered another round in my gun, and in a buckshot-fever panic ran down the road to where the buck had crossed. When I got there, I stepped up on the snow bank and looked down. The buck was standing right there, just below me in the deep snow, staring up at me, less than a stone's throw away. I realized then that he was a nice buck, a four-by-four with long brow tines. In America, they would call it a ten-point typical; in B.C. we call it a typical, four-point whitetail.

I up and aimed at him but this time he turned and ran into the trees below me before I could get off another shot. I jumped down the bank into snow up to my waist and started swimming my way through it, following his tracks. No sooner had I got through the thickest of the snow and into the trees than I heard a splash. *Oh no*, I thought. *He's jumped into the lake.*

I made my way down to the edge of the lake as quickly as I could and looked out. About 150 feet or so, out in the lake, I could see the big buck swimming away from shore, nothing but antlers sticking out of the water. I had to make a quick decision, to let him go or to take another shot before it was too late.

I looked down beside me, at the edge of the lake was a stump that had been cut off. I brushed all the snow off it with my hands and got down behind the stump in the snow, resting my gun across it. All I had for a target was a submerged head and antlers swimming directly away from me. This time, I took careful aim for the back of the head, stopped breathing, and squeezed. You could hear the thud as the bullet hit the buck in the back of the head.

The body slowly rolled over, floating lifeless in the freezing cold waters of Arrow Lake, 200 feet from shore.

Eight shots, pathetic, I berated myself. Now what am I going to do? My thoughts were interrupted by my partner yelling from up on the road. "What's going on?"

I yelled back up to him, "There's good news and there's bad news. Go back to the truck, get all the rope, bring my pack and all my clothes, and come down here."

He came down with everything a few minutes later, making his way through the trail the buck and I had furrowed in the deep snow.

"Well, what's going on?" he asked when he finally got down to me by the lake, huffing and puffing.

"Well, the good news is I got him. The bad news is he's in the lake. You see that clump of branches out there? That's him."

"What are we going to do now?" my partner asked.

By then I had formulated a half-baked plan. I told him about it. This was not going to be the first time I would have to swim to retrieve an animal. I was thinking back to a few years earlier when Keith had shot his first moose across the swamp and I had swum over to it, tied off the rope on its antlers, and swum back with it, across the slimy swamp. I remember how cold I had been during and after the moose retrieval and hardly relished having to do it again, but that was my plan.

I told my partner I was going to undress, tie the rope around my wrist, swim out to the deer, and wrap the rope around its antlers. Then he was going to pull for all his might, and get me back to shore as fast as possible. I would dry off and quickly change into my warm, wool clothes. I told him that if for some reason I didn't make it out to the deer and started going under, to pull me back in by my wrist, quickly!

He didn't like my plan. "You will drown or freeze to death out there; you can't go in the water!"

We stood and debated the issue for a few minutes.

During those few minutes, I noticed three things were changing. First, the light was running out and it was getting dark. Second, it was beginning to snow again. And third and most importantly, I noticed that the lake had some wave action happening—the water was actually lapping up on the beach and the movement of the lake was ever so slowly washing the buck back in our direction. My partner noticed it too.

"Look," he said. "The waves are bringing the deer toward us."

"Yes, I see," I replied. "But at this rate, the buck won't wash up on shore until tomorrow morning. We have to go get it now before its dark!"

"I have an idea," he said. "An old trick my uncle taught me."

"What's that?" I asked.

He went over and picked up the rope. Fortunately, I had brought a long length with me. My partner looked around on the beach for a piece of driftwood, a perfect piece, the size of a good chunk of firewood. He tied it on the end of the rope and asked me to hold the other end, and to stand back. He started twirling the rope around in a windmill-type fashion, letting out more and more so the motion got longer and higher each time. I realized what he was going to attempt to do, which was to hurl the driftwood out toward the deer, but I thought there was no way he could throw it that far. Round and round he twirled the rope. Then at the precise moment, he let fly with the driftwood tied on the end. To my absolute amazement, the piece of wood landed over top and on the far side of the deer floating in the lake.

"Wow, great shot," I said.

With great concentration, he slowly started pulling the rope toward the shore. Unbelievably, on that very first attempt, he worked the rope so the driftwood came across the deer and, by sheer luck which we needed about then, the piece of driftwood caught on the big buck's antlers and wedged in tight enough that he could begin pulling the deer in to shore. He managed to get the animal in close enough to where I could wade out with my boots and socks off and pull it all the way up onto the beach.

Cleaning it right beside the lake and in the snow was perfect; it kept it clean and neat, easy to dispose of the blood and waste. Upon careful inspection of the carcass, I realized where I had hit the deer and why it had turned down towards the lake instead of scampering up the hill and out of sight. My first bullet had been almost perfect but a little low. It had penetrated the buck through the rear and exploded internally. The deer had been mortally injured and, as is common with whitetails, it headed for the lake to escape.

I've seen deer swimming in the lake before, knowing they were pursued by cougars on shore. There was another shot in one of the legs as well, but as for the rest of the bullets, well, I couldn't find them or I had simply missed.

Once it was field-dressed, we gathered up our gear, grabbed an antler each and dragged the deer up onto the road. We walked back to the truck, put all our gear away, then drove the truck to where the deer was on the road and loaded it into the back. It was four o'clock on the last day of the hunting season.

"I think if we are lucky, we can catch the four-thirty ferry," I said. "Let's get out of here!"

We raced to the ferry with lots of time to spare. We pulled into the line-up, and with the truck running to keep us warm, closed our eyes for a well-deserved moment of rest. I had barely closed my eyes when all of a sudden, we were startled by someone banging on the driver's side window. It was a man in a uniform. The man identified himself as the conservation officer from

Nakusp, doing one last inspection for the year at the ferry before the end of the season.

"I see you have a deer in the back. Let me see your hunting license and tag," he demanded as he leaned in through the driver's side window.

My partner told him that I was the hunter and that I had shot the deer. He asked me to get out of the truck and show him my hunting license. I showed him my license and my whitetail tag which I had properly clipped. He looked at me and with a big grin on his face said, "Nice deer. Well done, gentleman!"

CHAPTER 7

Death Rapids

The Columbia River was once a mighty river. In 1964, the British Columbia and Washington State governments signed what was called the Columbia River Treaty and began building a network of dams on the Columbia River, over sixty in all, to supply power to the Pacific Northwest. Today, the Columbia River is a configuration of lakes, weirs, locks, and power dams.

The Columbia River had its headwaters in the Canal Flats area, south and east of Golden, B.C. From there it ran north to Mica Creek, swung south though B.C. into Washington State, and eventually flowed out into the Pacific Ocean at Astoria in Oregon. The Columbia River was a powerful river, with a high volume of water pushed through narrow mountain valleys—perfect for a grid of power dams.

I moved to Revelstoke just after the completion of the Mica Creek dam. Back then, the only natural section remaining of the

Columbia River was the 100-mile stretch between Revelstoke and Mica Creek.

I misspent many years of my life hunting and fishing north of Revelstoke on the mighty river, and don't regret a minute of it. I knew eventually the last of the power dams would be built at Revelstoke and the mighty Columbia would be gone. That is why I spent so much time appreciating what was left of the only remaining section of the river.

In the winter, I would drive up the dead-end highway to Mica Creek and watch the moose wintered along the river, pushed down by the massive snowfalls of the Selkirk Mountains. In the spring and summer, it was not uncommon to see woodland caribou on the road, as well as wolves, grizzlies, and black bears.

Access to the actual river from the highway was generally steep and challenging; fortunately, there were a number of private ferries that crossed the river to transport logging trucks and equipment to the west side of the Columbia. The general public was not permitted to use the ferries, but having them did provide roads down to the river from the highway, so I would drive down and park my truck at the ferry landings, then walk upstream from there to go fishing.

Fishing in the Columbia River was one of the best times of my life and I went often. I learnt quickly from friends and other fishermen how best to catch rainbow trout in the swift-flowing river. Back then, the river flowed from the Mica Creek dam down to the next dam, which was at Castlegar, well over two hundred miles away, so the variety of species of fish in the river were plentiful. Pea-mouthed Rocky Mountain whitefish; native Columbia River rainbow trout, distinguished by their yellow fins;

Gerrard rainbow trout, distinguished by their size and markings; Kokanee, which are a land-locked salmon; bull trout, which we improperly back then called Dolly Vardens; and the gentle giant, white sturgeon.

I targeted rainbows. Catching them in the fast-moving waters of the Columbia was a challenging and exhilarating feat. The river was silty and full of mica—fools' gold. The trick to catching the rainbows I learnt, was to use a small, gold-bladed spinner just above the hook and worm. Somehow the double-bladed gold spinner attracted more fish than without it. I would walk upstream until I found an outcropping in the river, where the current made a back eddy downstream of it. Walking out to the edge of the outcropping, I'd cast my line with a bobber attached, upstream. As the bobber was swept downstream, it hit the sweetwater spot where the eddy met the fast-flowing current, and with almost every cast the bobber would immediately dive under the surface and I'd have one on. Usually, I'd see them jump way out in the current, and the fight was on. It was so much fun. The rainbows all averaged between two and five pounds, although I have seen bigger caught, and ironically were gorged full of mice. I had many exciting fishing trips and adventures along the banks of the Columbia River.

I have a crazy white-water friend whom I've known for a very long time. We are like brothers. Gary is an experienced white-water canoeist; he has paddled many white-water rivers in Alberta and B.C. He was the inspiration for the idea I had of rafting down the Columbia River from Mica Creek to Revelstoke, before it would no longer be a river. The Revelstoke dam had already begun construction at Little Dallas canyon, I knew in a year or so there would be no more Columbia River left. The idea of "travelling" the last of the river consumed me day in and day out, so I began to research in detail the routes and hazards of the river.

Long before any dams had even been thought of or highways built, paddle-wheelers travelled upstream on the Columbia River from the south Kootenays and Washington State, to transport goods and people to the east and beyond. Many of them were American prospectors searching for gold. There were vivid accounts of a dangerous section of the Columbia River, forty-five miles north of Revelstoke that the pioneers referred to as "*rapides de le mort*"—Death Rapids.

I had been to the forestry office and picked up an old map of the river, but it did not specifically indicate where Death Rapids was. It did indicate four major sections of rapids between Mica Creek and Revelstoke. The first were at 80-mile, just below Mica Creek, called Priest Rapids; the next were at 60-mile (unnamed); then there were the rapids at 45-mile, with the last set being at 30-mile, the Downie Rapids. The only reference on the map regarding the rapids at 45-mile was a notation that read "unnavigable section of river."

When I was in the forestry office, one of the foresters asked me why I wanted the map, so I told him I intended to raft down the river. He looked at me as if I was crazy.

"You can't get through Death Rapids," he told me. "You will have to pull out at the 45-mile ferry, go around, and put back in at the 30-mile ferry."

"Why? Have you ever been through there or seen them?" I asked.

"No," he replied, "but I have heard they are unnavigable."

I had lots of friends who were loggers that had worked in the area of Death Rapids, so I queried all them about that portion of the river. All of them said the same thing: no, they hadn't seen the rapids but had heard they were certain death. "A waterfall," as one of my logger friends had told me.

Death Rapids is in a section of the river that cannot be seen from anywhere. They are in a steep canyon just around the corner and downstream of the 45-mile ferry, where the river boils through steep rock walls on either side of the canyon. Not one person I spoke to had ever seen them, but all feared them. The history books told the story of a mighty back-eddy at Death Rapids that swirled and sucked the under-powered paddlewheelers down to their death.

The forester who provided me with the map of the river offered some friendly but professional advice on the matter.

"Before you get to the rapids," he said "just past the 45-mile ferry, there is a bend in the river. Right after that, there is a large, calm eddy above the rapids. If you are uncertain at that point, pull over there and get out and walk downstream to see the rapids. If you change your mind about going through, you can always turn back and pull out at the 45-mile ferry."

Armed with only an old river map and the limited information of local residents, I called my friend Gary and explained the trip to him. Without hesitation, he agreed to go. Gary had an inflatable ten-foot Metzler rubber raft with a 7.5 horsepower Mercury outboard motor. The yellow raft had an inflatable bottom that was supposed to be better for running rivers than a hard-bottom raft. In hindsight, it wasn't much more than a blow-up raft that one buys for their children at a department store to float on a lake. It would be a gross understatement to say the Metzler was too small and under-powered to challenge the mighty Columbia River, but nonetheless we were determined to go.

I asked some friends to drive us up to Mica Creek, where we would begin our adventure. The waters below the dam were still and peaceful; they did not tell the real story of the power of the river. It was easy to gain access to the water there. One of the old roads that existed before the Mica Creek dam had been built now acted as a boat ramp. We easily put the raft in and pushed out to where it was deep enough to drop the leg of the outboard motor, and off we went.

Our plan was to get to Death Rapids that day and access the situation. Depending on time, we were either going to camp for the night above Death Rapids, at the 45-mile ferry landing, or just below the rapids. And crazy as it sounds, depending on the severity of the rapids, we planned to go back up through them and play around in them. We had no idea what we were in for.

Gary powered us out into the main current of the river.

Below the dam, the river was wide and fairly calm; the current did not seem to be any sort of issue for us.

We had noticed, however, that the dam had released a significant flow of water from the reservoir above it. The river was extremely high and full, with a lot of debris that had washed from the banks floating on top, creating a definite hazard for the diminutive inflatable raft.

I sat up front in the raft holding onto the rope that ran through the grommets around the top of the ridge of the raft; Gary sat in the back with his feet outstretched to my back, running the tiller, carefully avoiding any wood and debris in the river.

I had a picture of the map of the river etched in my brain by now. We soon came to the Priest Rapids, at 80-mile. The high water helped flatten out the severity of the standing waves, and with Gary expertly manoeuvring the tiny raft we went through the relatively easy rapids without incident or without taking on much water, although my baseball cap became a fatality when it blew off my head as soon as we hit the first waves.

"Yee-haw!" I heard Gary yell from behind me as we exited the last of the churning waters of Priest Rapids.

Cranking the motor to full speed, off we went downstream on our adventure. Rain had started coming down when we first set off and now it was a steady pour. Between the water already in the bottom of the raft and the rain, we were wet and going to remain so for the rest of the day.

Around 60-mile, there were some more standing waves we had to contend with, but again, the high water from the overflow of the reservoir kept the standing waves down to fast-moving humps and bumps; travelling through them was fun and easy.

Surprisingly, in fairly short order, we approached the 45-mile ferry area. I knew exactly where we were because I had walked upstream from the ferry many times on my fishing excursions. I pointed this out to Gary. We decided to pull over at the ferry landing and have lunch before heading down through Death Rapids. It was mid-afternoon by now and we were well ahead of our projected schedule. Time was not going to be a factor.

We had a quick bite to eat out of the rain under the shelter of the ferry, which was not in use at the time, but the apprehension of the impending rapids was palpable and hung in the air. We ate quickly and discussed our strategy.

We decided to take the advice given and pull over before the rapids and check them out.

We got back in the raft and slowly headed downstream to the blind corner waiting for us at the canyon. The river narrowed below the ferry landing and the current began to pick up speed; we had no idea what awaited us around the corner.

As we came around the corner, I was surprised to see that the river opened up, the current slowed down, and we were in a huge, calm area, like being on a small lake.

"This must be the calm eddy above the rapids!" I yelled back at Gary as the roar of the water grew louder and louder. "Pull over here somewhere."

Gary made a small circle outward then expertly turned upstream and set us ashore on the bank of the river, in the calm waters of the eddy. We tied up and walked downstream over the slippery rocks along the river. The roar of the river became deafening. We got far enough down to where we could see the beginning of the rapids, where the calm eddy spilled out into the main part of the river.

We were in a canyon, with steep walls on either side of the river. Here, we could see that the river made a hard left-hand turn, which we had just come around, into the canyon, and then, just beyond the calm eddy, made a sharp right turn, where the water hit the sheer rock wall of the canyon on the opposite side of the river. As the water crashed into the steep canyon wall, it created a set of standing waves that we estimated ranged from fifteen feet, decreasing downstream to a four-foot chop, a total of about twenty gigantic standing waves with water rushing over the top of them at an incredible speed, no match for our ten-foot Metzler. Between the standing waves on the opposite side from us and the shore on our side, was the massive, swirling back eddy, obviously the abyss of the legend that had swallowed up the early pioneer paddle-wheelers.

We had to yell at each other to be heard over the roaring noise of the white-water. "We don't want to be anywhere near this eddy!" Gary yelled.

"Well, I don't want to go anywhere near those waves over against that rock wall either!" I yelled back.

We looked further downstream and could see that the rapids ended, as the standing waves subsided and dissipated back into the river, a few hundred yards away, or so we thought.

"What do you think?" I asked Gary.

"I think if we stay out of the standing waves but keep to the middle of the river so we don't get sucked into the back eddy, we can make it through," Gary replied confidently.

"I'm game if you are," I yelled back.

Tentatively, we got back in the raft and shoved off out into the calm eddy. I remember feeling the same anxiety and adrenaline rush I get when I'm about to ski off a mountain peak or climb out onto an exposed ledge. It's a sickening yet exhilarating feeling. Over the years I have learnt how to use the adrenaline as a motivator.

Gary powered us out and into the mainstream of the eddy. Just as we were about to leave the calm of the giant eddy and head into the raging torrent of the river, I looked up, and hovering above us was a helicopter, the little bubble kind you used to see on ABC's *Wide World of Sports*. I pointed upward to Gary but the moment was fleeting. We were about to be swept into the current of the river.

Gary aimed the tiny raft into the middle of the gigantic, raging river but as soon as we hit the current, the river shot us straight toward the canyon wall, the 7.5-hp outboard was no match for the power of the river and became merely a steering mechanism, even at full throttle; straight for the first fifteen-foot standing wave we headed.

Miraculously, thanks to Gary's expert handling, we slipped off the side of the first standing wave, avoiding the angry white water crest at the top of the wave that would have undoubtedly flipped us over backwards, meanwhile keeping the deadly, swirling eddy well off to our right. Then, instantly, we were racing into the trough at the bottom of the next wave, and over it we went, again avoiding the dangerous breaking white water at the top, and then again, down and up and down and up as each wave became a little less intimidating until we were into waves without breaking white-water on the top of them, Gary now going right through the middle and over the top of each wave, yelling out behind me like a crazed boatman.

"We made it, we made it!" I yelled back in relief. "That wasn't so bad."

The river below us started to flatten out and became still and calm.

"What do you think? Do you want to go back up and try it again?" I asked Gary.

"Let's find a place to make camp first. It's too steep to pull over in here. We can unload the gear and then come back."

"Good idea," I concurred.

By now, the river was almost tranquil, like a lake on a summer's day. Gary had the raft in idle as we let the fast-moving current of the river carry us down, looking for a place to pull over to camp for the night. I noticed as I looked ahead downstream that, for as far as I could see, the river was completely flat and calm, in an eerie, unnatural sort of way. Then all of a sudden, out of nowhere, the river dropped out from underneath us, and at that moment we both realized we were going down over a waterfall and into a boiling cauldron of rocks and white-water.

As we went over the waterfall I looked up, and, unlike the mirage of calm water I had previously seen, I could only see angry white-water stretching out for the entire length of the river ahead of us.

We shot straight down into a terrifying trough of boiling white-water at the base of the waterfall, straight toward a massive rock directly in front of us. I was holding onto the rope on the sides of the raft for all I was worth. When we hit the rock, Gary throttled the motor up full blast, and by some stroke of luck, we went right over the rock and into the white-water rapids beyond, still upright!

Later, Gary told me that, when we hit the rock, I flew up so high he could see right underneath me, while I clung to the life-saving ropes of the raft.

Over and over we churned through the rapids, avoiding some rocks, sliding sideways or going completely over others, until after what seemed like an eternity, the river subsided, and we reached calmer waters further downstream. It was only then that I dared looked back at Gary, who was grinning at me.

"Pretty exciting, huh?" he said calmly.

"I thought we were done for on that first rock," I replied. "Great job on the tiller, buddy!"

I breathed a big sigh of relief.

Finding a spot to camp, we pulled over and got out, my knees still shaking. It was raining and we were completely soaked. We set up camp under a rock ledge, but everything we had was wet, except for one set of dry clothes that I had put away in a plastic bag. I changed into them and sat down for a well-deserved break from the river.

We couldn't get a fire started but I had brought my one-burner camp stove with us, so we heated some stew and warmed up. After supper and still with lots of light, Gary said, "Do you want to go back out?"

I said I did, so we jumped back in the raft and headed upstream into the boiling cauldron of white water again. It was actually easier going upstream than it had been coming down. I didn't want to put my soaking-wet life jacket back on again—I couldn't bear the thought of how cold and wet it would be, so I left it behind. We hadn't gone very far upstream when suddenly, we hit an unexpected rogue wave and out of the raft I popped, into the freezing cold water of the river. I was swept downstream in an instant. Gary swung the raft around downstream and skilfully came racing up beside me. I managed to quickly swim over to the raft and climb onboard, out of the river. So much for my only dry clothes. It wasn't fun anymore, so we went back to camp and spent a sleepless night in our wet clothes and sleeping bags.

The next morning it had finally stopped raining, so without consideration of going upstream anymore, we set off downstream for the not-too-far-off Downie Rapids. No sooner had we got back out into the main current again than the motor started sputtering and died. "We're out of gas," Gary pronounced. "Looks like we are paddling the rest of the way."

The Metzler, built with a transom to hold a motor at the stern and a pointed nose at the bow, is not exactly a paddling type of raft. Between the two of us, it was difficult to keep the raft pointed forward. The current spun us around and we found ourselves constantly floating backward downstream and trying to bring the raft around. On top of that, every time we went near an eddy, the current pulled us into it and we had to paddle hard to get out of it.

The Downie Rapids were tame compared to Death Rapids, and enjoyable because the current was swift through the rapids, which helped us pick up some speed, even if we did have to paddle through them to maintain our course.

We were expected back at a certain time, and without a motor we were not going to make it. We managed to float and paddle our way down the rest of the fairly calm river.

Our friends were anxiously awaiting our arrival. When we finally got down to the pull-out just above the Little Dallas Canyon, they were happy to see us, helped us pull the raft out, and drove us home.

A year later Gary came back with a twelve-foot Metzler that held five people, and a fifteen-horsepower motor. We went through Death Rapids five more times after that, before the river was finally flooded. On one occasion, I talked my sister into coming with us; she sat between Gary at the back and me in the front. To this day, she tells me it was the most terrifying thing she has ever done in her life!

CHAPTER 8

Nakimu Caves

I've had a number of different jobs over my lifespan, some good, some not so good. After I got out of high school, I moved to Revelstoke, B.C. While in Revelstoke, I became an accomplished climber, backcountry skier, mountaineer, outdoorsman, fisherman, and hunter. I fought fires for the Department of Forestry, drove snow-plough trucks for the Department of Highways, did search-and-rescue for the RCMP, and for two summers worked for Parks Canada. The two years I worked for Parks Canada in Mount Revelstoke and Glacier National Parks were my favourite by far.

I worked on one of the trail crews. If you get a chance to hike any of the numerous trails at Rogers Pass in Glacier National Park, don't let it go unnoticed how well-groomed and maintained the hiking trails are. That is not by accident. It is a direct result of dedicated park employees who work hard to ensure everyone stays safe and on the trails, in order to protect the Park's fragile ecosystem for generations to come.

I worked on a crew of six people, each of us with special abilities. During my second season, our crew had the very good fortune of being selected to go into the Nakimu Caves, led by a couple of the wardens.

The Nakimu Caves are a little-known set of caves in Glacier National Park, consisting of approximately six kilometres of tunnels and caverns. That might not sound like a lot, but the cave system is very extensive, and challenging at best. The caves were first discovered in the early 1900s and for many years, long before the Trans-Canada Highway ran through Rogers Pass, spelunkers and tourists alike from all over the world travelled by train to Glacier House, a hotel at the foot of Rogers Pass, to hike up into the caves.

Today, for the protection of the caves' fragile ecosystem, they are closed to the public.

Also fortunately for me, our group chosen to go into the caves was led by one of the most experienced wardens in the park. I knew him well; we were friends and had been on a couple of climbs and a rescue mission together. We both respected each other's ability. The wardens assessed our crew's ability as "experienced" and therefore decided to enter the caves from an alternative and more dangerous, seldom-used route, which involved rappelling down a waterfall into a large cavern called the Queen's Ballroom.

The first warden rappelled down and secured the bottom line for the rest of us. I went second, with little hesitation. I remember the ominous feeling of rappelling into complete darkness, except for the light from my headlamp and the guiding, dim glow of the warden's headlamp far below me. The waterfall itself was merely a trickle at the time, but there was still enough flow that I had to try and avoid the main stream as I descended into the darkness.

Once I was safely on the cavern floor, I had a chance to look around the massive cave, the largest in the Nakimu system. Gigantic, broken pieces of rock hung from the ceiling, waiting to fall to the ground, just like all the other ones that had fallen and that I now found myself having to climb over. I remember thinking that if one of those rocks let go from above, it would squash me like a bug. Then I put the thought out of my mind—because I had to. It took tremendous mental stamina to stand in complete darkness, not knowing where you were, with the looming threat of sudden death just a mere earth-tremor away.

The Nakimu Caves are formed by a river which bores its way through a soft, sedimentary layer of limestone within the solid rock formation of the mountain. We call it the "River of Caves." Over many years, the river has changed course, leaving behind a labyrinth of dry tunnels and caves.

Once everyone was safely down, we began exploring the more commonly known tunnels. The tunnels are incredibly smooth and perfectly rounded from the millennial years of the river carving its way through the soft limestone.

One of the unforeseen dangers is going down a tunnel without tying off a rope at the top. It's easy to get excited while you are exploring the caves and slide down a smooth tunnel, but almost impossible to crawl your way back up it without aid; like trying to climb up a steep, twisting water-slide from the bottom. I soon realized this and decided to follow closely behind my friend, the warden.

There existed a small, narrow passage that had never been explored before by any of the Park's members. Every time the wardens go into the caves, they are tasked with mapping them, to provide a better understanding of the elaborate tunnel system. Our expedition was no different. We were tasked to explore a narrow passage that no one had been through before.

The wardens gave us all a choice to follow them through the passage or wait at the entrance until they returned. Most of our crew decided to wait, but not me. I went second, behind my warden friend. The tunnel was so small that we had to start off on our stomachs and worm our way through. At one point, it became so confined that we had to remove our helmets and lights and reach out in front of us with them in hand, with the top of the tunnel scraping our backs. I would never do it again, but at the time I had no choice but to keep moving forward through the tiny passage.

I could constantly hear the roar of the river boring its way through another tunnel above us. I remember thinking, *Just the slightest shift in the earth anywhere nearby and I would be squished or trapped*. But then I put the thought out of my mind, as I said, because I had to.

Eventually the tunnel opened up into a small cavern, big enough that we could stand up, crouching over. From that cavern, there was an alternate exit route known by the wardens; we were not going to have to go back the same way, to my great relief. Eventually, once we met up with the group, we exited the caves through a large opening just above the Queen's Ballroom, a much easier egress point than the way we had first entered.

Many years later, after I had moved away from Revelstoke and was married, I started thinking about the caves again. In one of my less-brilliant but spontaneous moments, I talked my wife into going for a hike up into Glacier National Park, and, as promised, a tour of the caves. We invited my wife's sister and her boyfriend along for the hike as well, both fit athletes.

The day started off bad and got worse.

Everyone climbing in the National Parks must register in and out at the warden's office. I knew we were not allowed to be in the caves, so I did not go to the warden's office to register. The trail head to hike up to the caves is not easily discernible from the highway, but of course I knew exactly where it was.

As we began the five-mile hike up to the caves, it started to rain lightly. We hadn't gone more than a few hundred yards from the highway when we came to some bright yellow flagging tape stretched across the trail, the kind you see when the cops tape off a crime scene, warning of grizzly bears in the area.

"*Do not proceed, turn back,*" it read.

We had driven a long way to get here, and not being one to be deterred easily, I convinced "the crew" that the wardens only put the flagging tape up to discourage people from hiking up the trail because they don't want anyone going into the caves. I managed to convince everyone to carry on, so upwards we hiked.

The seldom-used trail going up to the caves is actually prime grizzly-bear habitat. The fresh piles of grizzly bear scat along the trail were evidence of that. With each fresh pile we came across, it took more and more convincing to keep everyone going; the girls wanted to turn back. We had a theory that we always believed on the trail crew: that there has never been a reported attack by a grizzly bear on a group of people where there were six or more people in the group. Of course, our little group consisted of four, but I convinced everybody to stick together and make as much noise as possible. Reluctantly, we pushed onward and upward.

Once we broke out of the trees and into the alpine, the rain became more noticeable. The river that is the origin of the Nakimu Caves, flows from above the caves, goes underground through them, then exits above ground further down. We came to the river below the caves, where the trail crosses it. The water was low enough that we could hop from rock to rock to get across without getting wet; mind you, by this time, we were already pretty wet from the rain.

We finally arrived at the entrance to the caves. Most of the obvious entrances were barred off, like you would expect to see in a jail cell, and bolted into the rock. I knew where one of the entrances was that was not sealed off. It also gave us cover from the rain, so we huddled underneath the rocky ledge that covered the entrance to the cave and changed into dry clothing. I had brought coveralls for everyone and a headlamp for going into the caves. Once we had all changed and geared up, I gave a brief safety talk. I decided to rethink about going into the depths of the cave. I suggested instead we just go in as far as the entrance to the Queen's Ballroom, which would be below us, and stop there. I really just wanted everyone to get a feel of what the caves were like without having to do anything too dangerous.

I led us to the brink of the big cavern, and as best I could, shone my light down into the abyss of the Ballroom. I told my uneasy crew that we weren't going any further, I just wanted them to experience complete darkness, so I asked everyone to turn off their headlamps. Until you have stood underground with your eyes wide open, you have not experienced complete darkness. With our lights out, we could not even make out the persons standing beside or across from us.

At the very same moment we were standing there—in the complete darkness—I thought I heard voices.

"Shush," I whispered "Listen."

Sure enough, from below us I could hear voices, then I saw three tiny lights. Obviously someone else was in the caves too. I really couldn't figure out who these people were or where they had come from because when we parked on the side of the highway at the trail head, there were no other vehicles.

I switched my light back on so the strangers could see we were in the caves as well, yelled out to them that we were there, then told my crew that we should head out, which we did. The four of us again huddled at the protected entrance outside the cave, changed out of our coveralls, put all the equipment back into the packs, and grubbed up in preparation for the climb down. My trail mix is legendary—everyone dug in.

While we were huddled under the ledge prior to our retreat, the three cavers came out. To my surprise, they were three Park wardens; one of them I knew from my trail crew days. I said hello to them as they abruptly passed by. Only the one I knew responded with a sheepish hello, then off they went down the trail, ahead of us.

We too gathered our packs, donned our rain gear, and began down. By then it was pouring rain. When we came to the part of the trail where the river crossed, we were astonished to see that the little stream we had hopped across on our way up was now a torrent of raging mountain water. It was way too severe to try and cross. We would have been swept away and down the mountain if we tried—and we had to get across to get down.

I dug out my climbing rope. I told my sister-in-law's boyfriend that I was going to tie the rope around me and go across; he was to hang onto the other end of it, and if I slipped or got swept away, he was to pull me back.

I further instructed him to stay where he was once I got across. I would then hold the rope from the other side while he held it from his side, so that the girls could hold onto it like a clothesline, to get across. Then he was to tie the rope around himself and I would keep a watchful eye on him as he came across.

As precarious as the river crossing was, it went as planned and without a hitch. We all managed to get across safely—soaking wet, but safe!

We hiked the rest of the way down without incident and came out at the highway. The three wardens must have been just ahead of us: they were loading up their vehicle and getting in when we arrived. The one I knew came over and greeted me, away from the other two. He said he hadn't meant to disrespect me up on the mountain, it was just that the other two wardens were pissed off that we were there. We spoke our platitudes quickly and said goodbye.

Months passed, and the memory of the hike into the Nakimu Caves waned—that is, until one day, about six months later, when I was sitting in my office at work and a Park warden walked in. I knew the warden, a friendly local man I sometimes drank with at the pub, someone I had met during my earlier days working on the trail crew, but not one of the wardens I had encountered at the Nakimu Caves.

I greeted him and asked what I could do for him. Sheepishly, he said he had come to arrest me. I laughed and asked him what he was arresting me for?

He said, "For being illegally in a cave."

I thought he was joking, but he said it wasn't a joking matter, that one of the wardens who'd seen me up at the caves was a hardass and wanted to make an example of me. Then he apologized, handed me the summons, and left my office.

On the day of my trial, I waited anxiously out in the gallery of the courthouse with all the other hardened criminals.

Surprisingly and unbeknownst to me, my lawyer walked in, came up to me, and greeted me. He was my personal lawyer, my corporate lawyer, and a man I'd invited to my wedding. On many occasions I had played squash against him.

I asked him what he was doing there. He said that, when called upon, he acted as the prosecutor for the Federal government; that, since I had been charged by Parks Canada, the charge fell under federal jurisdiction; and that he would be prosecuting me for illegally entering a cave in a National Park.

He said, "I suppose you are going to plead not guilty," knowing my independent nature, or perhaps trying to give me a way out.

I told him, "No, I am going to plead guilty and ask for leniency from the court."

When my case came up, I stood in front of the judge, next to my lawyer, the federal prosecutor—at this point I was thinking federal traitor!

My lawyer read out the charges to the judge: "That I did unlawfully and willingly enter a cave in a National Park that I knowingly or ought to have known was prohibited from the public."

The judge looked at me and asked, "Is that true, did you do that?"

I told him, "Yes, I did, your honour."

I then went on to explain that I had only entered the cave to get out of the rain, and to dry off and change my clothes. I told the judge that I was sorry.

Puzzled, the judge then asked my lawyer what kind of a penalty went along with such a charge. My mouth almost dropped when my so-called friend, the federal prosecutor, said it was a two-thousand-dollar penalty and up to five years' imprisonment.

The judge pondered for a minute, and then, with a wry smile, delivered his sentence: "Twenty-five-dollar fine. Do you need time to pay?"

I thanked him and assured him I would pay it on my way out. I could see the judge trying to compose himself, trying not to look like he was making a mockery of it all.

The next week, in the local newspaper, the front page headline read: "Cave Dweller Nabbed," with an edited version of the incident and my name splashed all over the article.

Shortly thereafter, I received calls from all my climbing buddies, offering to pay me twenty-five dollars to take them into the caves.

CHAPTER 9

Eagle

I think one of the most beautiful parts of Canada is between Glacier National Park and Banff National Park. Most tourists and travellers think they are travelling through the Rocky Mountains when they drive through Glacier National Park and Rogers Pass, but in fact they are driving through the hauntingly beautiful and desolate Selkirk Mountains, a completely distinct and separate range of mountains, much older than the better known Rocky Mountains.

Rogers Pass is steeped in history and plays a significant role in the transportation of goods across Canada; it is the corridor to the Pacific Ocean from the east. At the summit of the Rogers Pass is the interpretive centre for Glacier National Park, as well as the warden's station for the park.

Glacier National Park is one of the oldest parks in Canada, conserved for its beautiful, rugged peaks and multitude of glaciers.

The Canadian Pacific railway company first pioneered the route through Rogers Pass in the early 1900s. Then, in 1965, the Trans-Canada highway was completed through the pass and is today the main route from Calgary to Vancouver.

When the railway line first went through Rogers Pass, prior to the Trans-Canada highway, the stunningly beautiful Asulkan Glacier came to the base of the pass, currently where the highway runs; however, that was a hundred years ago. Today, the Asulkan Glacier has receded to a mere portion of what it was then and can only be reached by hiking up one of the many trails in the Asulkan Valley. This is where my story begins.

As you drive up the Trans-Canada highway and enter the last long, sweeping bend before the ascent up to Rogers Pass, from the west heading east, you can't help but be overwhelmed by the magnificence of the four major peaks that are prominently towering above the highway. To most travellers, they are just mountains, but to me they have as much reverence as the Himalayas do to the Nepalese Sherpas.

The four peaks, from left to right, are Avalanche Crest, Eagle, Uto, and Sir Donald. I have climbed three of the four; unfortunately I never had an opportunity or a weather window to climb Sir Donald, which is considered the Everest of Glacier National Park. I console myself in knowing that Uto, beside Sir Donald, is as technically difficult to climb as Sir Donald, just not as high or as long.

My father was born April 22, 1926. He first moved to Revelstoke in 1972, when he and my mother purchased a motel and campground there. Dad fell in love with the mountains and stayed there until his passing. He was a religious man and never ceased to marvel at the beauty of the mountains God had created.

I worked in Glacier National Park for Parks Canada. It was during that time that I learnt the *"lessons of the mountains"* and how to become a mountaineer. I use the term mountaineer and not climber because the immediate impression of a climber, for most people, is someone you see scaling a rock face with pitons and ropes. Mountaineering encompasses all aspects of climbing, from hiking rugged, valley bottom terrain, to route finding through the trees, to crossing open snow fields and glaciers, and then eventually free-climbing rock faces in order to reach the peaks of the mountains.

I rarely stayed overnight on a mountain; most of my climbs were one-day climbs, up to the top and back the same day. Sometimes that meant leaving at three a.m. and having to walk through the valley bottoms and timber in the dark. My one overriding rule was to always establish a turnaround time, usually four p.m. in the summer when it stayed light until eleven, and one p.m. in the winter, when it got dark earlier. I hold fast to my turnaround-time rule, sometimes agonizingly being five hundred metres from "peaking it," only to have to turn back. Five hundred metres from the top may only be an hour or so away, but it is also an hour back, provided everything goes well! It is not worth the chance, and I especially hold true to this rule if I am climbing with anyone who is not as fit or experienced as me. Generally, it takes just as long to climb down as it does to climb up.

I would tell my dad about each mountain I had peaked, and he would hang off every word I said and wished he had been there with me.

Dad wanted to climb Mt. Begbie in the worst way. "Begbie," as it is known, is a magnificent mountain that is visible from downtown Revelstoke, an iconic landmark, coated in glaciers and snow most of the year. It is a popular climb for budding enthusiasts but yet not one to be underestimated. I have climbed it twelve times—twice on search-and-rescue missions, a story for another time.

Begbie is not in any park; as a matter of fact, it is not even in the Selkirks. *Begbie* lies just west of the Columbia River across from Revelstoke in the Monashee mountains, a lesser range than the Selkirks.

I told Dad that I would go with him, but he needed to get in better shape if he wanted to climb Mt. Begbie. He was sixty-two at the time, and it wasn't that he was in bad shape—I just didn't feel he was in good enough shape to climb Begbie. It is a twelve-hour day for most hikers. In the alternative, many climbers overnight at the foot of the glacier on Mt. Begbie, then climb to the peak the next day and back down; but for me this meant carrying a bigger pack with more equipment, and that is exactly why I like to get up and down in one day. So Dad put his bike away and started walking everywhere instead.

The following summer, Dad, along with two other much younger men, one being an RCMP officer, and I, climbed to the top of Mt. Begbie. It was for him a crowning achievement. I was extremely proud of him.

Now that Dad thought he was Sir Edmund Hillary, he wanted to climb one of the peaks in Glacier National Park at Rogers Pass, a more formidable task.

There are many trails in Mt. Revelstoke and Glacier National Parks. You do not have to be a mountaineer to enjoy the scenery that the vast network of trails offers, but if you want to climb off the trails to the top of one of the mountains, well, that is a different story.

The first thing you have to do is register in and out with the wardens' office and specifically tell them where you are climbing and when you will return. Aside from that, you need all the right equipment, including crampons and an ice axe for crossing glaciers.

I knew of a fairly easy peak in Glacier National Park that we could climb, one without any glaciers and without a lot of exposure, so the following year I asked Dad if he wanted to climb Eagle Mountain, one of the four main peaks leading up to Rogers Pass. Without hesitation he said yes!

So the following year we made it our goal to climb Eagle. We drove up to Rogers Pass and, as required, registered in at the wardens' office. I knew almost everyone there from my previous years of working in the Park; I was good friends with the radio operator, whom I had worked with before on the trail crew. He was now the go-to person for registering in and out of Glacier National Park.

Dad and I set off from the Asulkan parking lot and followed the trail up, toward Avalanche Crest, a long but popular hiking trail. Midway up the trail we came to an area that some of us affectionately refer to as the *teahouse*. From there, you can look up and see the visible peak of Eagle and the iconic rock formation that looks like an eagle overhanging the mountain, the derivative for its name.

At the teahouse, we left the well-groomed trail to Avalanche Crest behind us and started up to the bottom of Eagle, beating our way through the brush—rhododendron, devil's club, and relentless slide alders. Staying along a small watercourse, the brush eventually gave way to large boulders and talus rock. I don't know if talus is the proper word for it or not, but it is the one I picked up from other climbers, for the rock scree that falls off mountains, the habitat of hoary marmots and tiny mountain pikas (rock rabbits).

Eventually, after scrambling up over the endless rock scree, we came to the steep slope that led up to the southwest ridge of Eagle. The slope was covered in small balsam trees which made precarious handholds, as we pulled our way up and onto the ridge, then followed the ridge up to the peak.

Comparatively, it is fairly flat on top of Eagle; the views, however, are as compelling as those of any of the larger, neighbouring peaks—truly magnificent!

Whenever I look out from any of its peaks, I'm always humbled by the vastness and extreme remoteness of the Selkirk Mountains. Dad, he didn't say much once we were on the top of Eagle. He just sat there, exhausted, drinking up the glory of it all. The sun was shining, it was a wonderful day.

As with most major peaks throughout the Park, there was a cairn set up on the top of Eagle from previous climbers, each climber adding their one rock to the pile. I let Dad place our ceremonial rock on the cairn. He also took out a small piece of paper and wrote his name on it, along with a prayer and the date, placed it in a plastic film container, and buried it within the cairn.

Before heading back down, I went over my regular safety talk with him, just as a reminder to keep focused on the task at hand, which was to get down safely. I reminded him that climbing down is more dangerous than climbing up and that we still had half as far to go yet. Also, when descending Eagle, great concern needs to be taken so as not to roll rocks down on lower climbers; it is a steep descent with loose rocks.

We worked our way down off the solid safety of the rock at the peak until we came to what is the steepest and most precarious part of the descent, an alternative route from the one we had taken up. We had been walking fewer than thirty minutes since our break at the top but regardless, I sat Dad down and made him drink some more water and eat a few more handfuls of trail mix. I again went over the safety drill, especially the one about rolling rocks down on each other.

As we descended off the solid rock face of Eagle, we began down a very steep and narrow gully of dirt and loose rock. I explained to him that I would take the lead down, as I had coming up. I would traverse across the gully from one side to the other at a downward angle, then when I had reached the opposite side of the gully, I would stop and wait for him to come. He was to follow my exact path of descent over to where I was, and then I would continue back down and across to the other side of the gully, zig-zagging our way down, one at a time, until we reached a wider and less steep portion of the gully. In this way, if we did accidentally roll a rock down the gully, no one would be below it in its path.

I made the first descent down and across the steep gully safely, and once across, motioned for Dad to start down. About halfway across the route I had chosen, there was a small outcropping of rocks; I had made sure to cross above them. I watched Dad come down across toward me and noticed as he got to the outcropping of rocks, he had come to it at a slightly lower angle than I had. He hesitated for a moment, considering if he should go down underneath the outcropping, or climb back up the gully to get around it. It was definitely too steep to descend underneath the outcropping, so he laboured back up the slope to cross above it, as I had.

He was tired, as any sixty-three-year-old would be after climbing all day with a fit thirty-year-old. As he lifted his leg to step over the outcropping of rocks, he tripped as his front foot stubbed one of the rocks and immediately fell forward. The slope was so steep that his fall forward was more like a free fall from a diving board; he literally flew over my head as he fell down the mountain.

I was standing in a precarious position as it was, realizing the steepness of the slope below me; nonetheless, I tried to reach out for him as best I could, hoping I could grab his pack or knock him down somehow, but it was in vain. The first time he hit the ground was hundreds of feet below and in front of me, then I saw him bounce off the slope two or three times again, and over a precipice and downward, out of sight. I knew it wasn't good.

The first thing I said to myself was *be careful, don't rush*. I urgently but cautiously descended, zig-zagging my way down and across the steep gully, looking over every little edge to see if I could see him, but I couldn't.

I first found his day pack and gathered it up, then I noticed other various items that belonged to him as I traversed my way down the gully. Finally, when I got to the bottom of the gully and started seeing patches of grass where the slope became less steep, I saw his body lying motionless below me on the valley floor. I estimated from where he had tripped to where he was lying, it must have been over one thousand feet vertically. I immediately realized once I saw where he was, that there was no way he could have survived the fall!

I continued down to where he lay. As I walked up to him, a flurry of mixed emotions ran through my mind. *What if he's still alive? How do I verify that he is dead? How am I going to get him out of here?*

When I walked up to him, he was lying face down, completely at peace in a little patch of green grass, the only such grass in the rocky area. As I approached him, it seemed there was an aura of soft yellow light emanating from and around his body, like fluorescence on the ocean; it stopped me. I just stood there for a minute or so. Then, as I watched, the light began rising up from his body; like mist on a warming morning, it turned from yellow to green, then blue, and evaporated up into the sky. I knew then that he was dead!

I checked his wrist and then his carotid artery but could sense no pulse from either, no sign of life. I assessed that he must be deceased.

One of the articles of clothing I discovered in his pack was a reversible coloured vest, one side green, the other fluorescent orange. I placed the vest, orange side up next to him, with a couple of rocks on it to hold it down, in hopes it would be visible from the air, and decided to hustle back to get to the warden's office at the Rogers Pass centre before it got any later, or darker.

When I walked into the warden's office, my friend the radio man said, "Thank goodness you're back, what took you so long. Is everything okay?"

I said, "No, my dad fell to his death on Eagle!"

At first he laughed, thinking I was having one over on him, but then he realized from the look on my face that I was serious.

Grasping his hands over his mouth, he said, "No way!"

I told him where my dad was. He immediately called for a helicopter. Two wardens were assigned to the helicopter, both of whom I knew well, one I had been with on a rescue mission on Mt. Begbie the year prior. They asked me to come with them in the helicopter to show them where my dad was.

The pilot flew the three of us up and over to Eagle. I showed them the gully we had descended; it was quite easy to see Dad's body lying next to the bright orange vest at the base of the gully from the air. I thought we were going to land and recover the body but they instructed me that they were taking me back to the pass and dropping me off, then coming back for him.

I drove the forty-five miles home to Revelstoke in a complete daze, not feeling any emotions, numb, not knowing what or how to feel. I found my mum in the busy campground; it was the August long weekend. I took her aside and began to tell her what had happened. It was then I burst out crying. She just hugged me to console me and said, "Don't cry, my son."

There were many side stories that followed; my dad and I were well known in Revelstoke. Everyone felt sorry for me, as you can imagine. Ironically, I didn't feel as bad as everyone thought I must. In later years, I have consoled myself with a number of wonderful memories. If Dad could have written the script for his ending, he would have wanted to die exactly as he did, high up in the Selkirk Mountains on a bright, sunny day.

I'm not one either to see visions or believe in the occult—far from it, actually—but when I saw the light that radiated from

Dad's body for that brief moment, I believed his soul had been delivered; somehow it gave me closure.

Later the next year, one of the senior wardens, whom I had climbed with a few times in the Park, called me and said that he had been tasked with collecting all the memorabilia from the various cairns on the different peaks throughout Glacier National Park. Parks Canada wanted to gather the artifacts before they got destroyed by ravaging winter storms and avalanches.

He told me he had collected and read my father's note from the cairn on Eagle and had requested it be placed on display in the interpretive centre of Glacier National Park.

CHAPTER 10

Albert Peaks

The Albert Peaks are an iconic set of peaks in the Selkirk Mountains, consisting of two main peaks, the North and the South Albert Peaks. They lie directly east of the city of Revelstoke and, from certain locations in Revelstoke, are stunningly visible, especially in the winter, with their snow-clad tops.

The Albert Peaks aren't a go-to climbing destination for most climbers; however, it was always my ambition to climb all the mountains that were visible from Revelstoke, and that included the Albert Peaks.

Very few climbers that I know have ever climbed the Albert Peaks—or wanted to, for that matter. The two biggest obstacles in climbing the peaks are the fact that the rock is rotten—that is to say, the rock is unstable, crumbly and easily broken, like shale, with a build-up of overburden on it—and the fact that the Albert Peaks are hard to access because the base of the mountain lies beyond the main Canadian Pacific rail line and on the other side of the Illecillewaet River (pronounced ill-ee-sill-ee-what), making it an all-but-impossible hike, without any discernible trails, to get to them.

Fortunately in 1985, a logging road was built that went across the CPR railway tracks at Greely, and a bridge put in to cross the Illecillewaet River, making access to the Albert Peaks a lot easier. Today, both those access points have been removed and the Peaks are seldom considered worth climbing.

My climbing mentor and the person who got me started climbing real mountains in the Selkirks was my good friend Jon. I call him Spiderman; he could climb anywhere and anything. For fun, he would climb the sides of buildings, using only the mortar gaps between the bricks for handholds. Often, Jon would ask me to take pictures of him out on a ledge when we were climbing together; I was always too scared to even look over to where he was and refused to take his picture. Jon was younger than me and had more experience as a mountaineer than I did, but our fitness levels were comparable, so we made a good climbing team.

The first winter the logging road went into the Albert Peaks, another friend of mine, Christian Maier, called to let me know there was a logging road open to the base of the Albert Peaks, and that we should go in and ski some of the backcountry slopes. Backcountry slopes often meant skiing avalanche chutes, a tenuous and precarious undertaking, not one that should ever be done without vast experience and a complete understanding of weather and snow conditions!

By day Christian was a logger, so he knew where every new logging road was. But his real passion was backcountry skiing, and so on the weekends, when the logging trucks weren't on the winter roads, we would drive up to the top landings of the logging roads and tour into many of the inaccessible backcountry peaks. These are some of my most impressionable memories.

Many considered Christian to be the best backcountry skier in Revelstoke; I considered him to be one of the best backcountry skiers in the world at the time. Sadly, Christian's life was cut short at forty years old when he was killed in a logging accident.

After that first winter of skiing the Albert Peaks with Christian, I told Jon about it, that there was a road to access the mountain. So the following summer, we made a plan to climb the North Albert Peak. I wanted to climb the north peak because I'd been fascinated by it for years, driving up to and from Rogers Pass. On many occasions I had stopped on the side of the highway to view the goats living on the north peak from across the other side of the Illecillewaet, and while observing the goats had wondered about climbing the north peak and being able to look back down onto the highway from the top. The south peak did not overlook the highway.

So later that next summer, Jon and I set off early one morning to climb the North Peak. We drove as far as we could up the logging road, parked, then worked our way through the bush by traversing a recent burn, and eventually made our way out of the trees and onto the talus below the north peak. We chose to do a ridge climb instead of the shorter but harder face climb. As it turned out, the ridge climb was precarious, with loose rock all the way up. Every step or handhold had to be double-checked before trusting it.

We reached our determined turnaround time of four p.m. and still hadn't reached the top of the North Peak. With the ridge getting narrower and the rock becoming less stable, we decided to turn back and call it a day. I left disappointed and amazed at how difficult and how far it was to the top of the North Peak.

In the spring of 1990, the year after my dad had fallen to his death on Eagle, Jon called me up and said that I needed to go on another climb—"to get back on the horse," as he put it.

We talked about various mountains we both wanted to climb but ended up agreeing on climbing the South Albert Peak. At more than ten thousand feet, the south peak was actually higher than the north, and we were still disappointed that we hadn't made it to the top of the north peak previously.

Ten thousand feet may not seem like a big mountain to some, but for the Selkirks, it is up there. I had never previously climbed over ten thousand feet in the Selkirks; most of the mountains are in the nine-thousand-foot range.

You have to realize that we start climbing from the valley bottom to the top and back in one day. The valley bottoms are about fifteen hundred feet above sea level, so the vertical up and down on a one-day climb is over fifteen thousand feet. And it is always the last thousand-foot ascent and the first thousand-foot descent from the top that take the longest and are the most dangerous. Hand grip, foot grip, hand grip, foot grip—then move one grip while maintaining a three-point stance at all times. It takes a long time to climb a mountain that way!

So on a nice, warm, and sunny morning five years after climbing the north peak, Jon and I set out for the top of the South Albert Peak. We parked at a place on the road where a small stream came off the mountain between the two peaks and made its way down to the Illecillewaet River. The stream was overhung with slide alder but still offered us the most direct route through the trees up to the base of the south peak.

It was early spring and the run-off hadn't started yet. We managed to work our way up the sides of the relatively dry stream bed through the rocky boulders and came out below the base of the two Albert Peaks into the alpine, above tree line.

We stood in awe of the magnificence of the mountain as we stopped for a drink of water and to replenish our bottles from the stream. One important aspect of mountaineering is to make sure you refill your water bottles at every opportunity possible, especially before climbing onto the rock face. Once you get on the mountain, there is nowhere else to get water, and the exertion and adrenaline from climbing is exhausting and dehydrating. For this climb, water conservation was going to be paramount.

Our route was fairly well defined. The south peak was an obvious face climb. We would have to scale the west face to the northwest ridge of the South Albert Peak, and then ridge-climb to the top.

The little stream we had been following began its journey at the tail end of a glacier that lay in the gap between the two peaks. The glacier swept upward to the steep rock face of the mountain. I had never seen the glacier before and didn't know it existed until then, but I always carry an ice axe, and the lily-white snow covering the glacier was still winter-thick, enough to hide the crevasses below, so crossing the glacier up to the rock would not pose that much of a challenge.

We unstrapped our ice axes from our packs and began the slippery trek up the glacier to the rock face. At the top of the glacier, where the ice and snow meet the rock face of the mountain, there is always a moat. A moat is the gap between the glacier and the rock face, where the glacier has receded away from the mountain. Great care has to be taken to cross from the glacier onto the rock. The moat at the base of the South Peak was deep. If you fell in, there would be little hope of a rescue, so walking up to the edge of the top of the glacier was a precarious task.

You have to expect that the edge of the glacier will break away and drop you into the moat. It is with this precaution I worked my way up to the edge of the glacier at the base of the mountain, making footsteps like rungs on a ladder for Jon to follow, while he held a short rope tied to me in the event the glacier edge gave way. Once I got to the highest point, where the glacier was closest to the rock, I packed the snow down with my feet to make a stable ledge, and leaning forward over the four-foot-wide gap, reached out for the solid rock of the mountain. Once I had straddled the moat, I pulled myself up and onto the rock. Jon came up and did the same.

Jon usually led our climbing expeditions; this one was going to be no different. The one thing I appreciated about the way Jon climbed was that he always assessed each route and then asked me if I felt comfortable taking it.

Following Jon up, we soon realized that the rock on the south peak was just as rotten as on the north. With every new handhold we reached for, we had to pull down hard to double check to make sure it wouldn't break off before putting our full weight on it. Most of the rock also had overburden on it, loose debris that had fallen from above, making it a long, slow, careful climb.

I followed Jon up the rock, one precarious hold at a time, until after some time we came to a small ledge, big enough for both of us to stand on beside each other. We had come up at an angle away from the glacier; it had seemed fairly easy so far. Looking back down the way we had just come up, I made a visual note of the route so that if we were to come back the same way, I would remember how to get down again. Route finding is not just looking forward, it is also remembering to look backward.

Jon asked me how I was feeling. I told him I felt good, so we pushed onward and upward in an opposite angle, towards the northwest ridge.

A little further up the face we came to an overhanging ledge. Overhanging ledges scare the crap out of me. I hate being exposed, leaning out backwards to climb around a rock outcropping.

Jon looked back at me and said, "Are you comfortable going around this way?"

I told him I wasn't. I looked up above me and saw a narrow chute full of snow that looked like it would take me up to the ridge.

"I'll go up here," I told Jon, "You go around."

The steep chute looked easy enough to begin with, handholds on each side and snow for foothold grips underneath. I started up it while Jon disappeared out of sight, around the overhang. I looked up and could see the top of the ridge, but the chute narrowed near the top and the rock handholds disappeared. The snow I had been using for footholds also disappeared as the chute got steeper and steeper, the snow now turned into a sheer wall of blue ice.

I used my ice axe to gain a few more feet but then ran out of options. I was dug into the blue ice with my ice axe, hanging off the steepest part of the chute, with the ridge just above me, but I couldn't make another move. If I were to start sliding backward, I wouldn't be able to self-arrest and there was no doubt I would slide down the chute and bounce off the rocks to the bottom of the mountain. I was stuck, hanging by my ice axe, unable to go up any higher and unable to go back down.

I called out for Jon. After a few minutes, I heard him from above: "You looking for me?"

Jon had managed to climb up to the ridge and was looking down the chute at me.

"I'm stuck," I said, trying to hide my fear.

"I'll lower the rope down for you," he replied.

Jon unstrapped the climbing rope from his backpack and lowered it. As he did, loose gravel and small rocks showered down on me. I had to close my eyes. A few larger rocks dislodged and I had to deflect them from hitting me with my one free hand. Jon saw what was happening and began lowering the rope as slowly and carefully as he could, without causing any more disturbances, but in doing so, the rope got hung up on a clump of ice just above me.

"Can you reach the rope from there?" he yelled down. "I don't want to pull it back up and drop anything more on you!"

I opened my eyes and saw the rope above me, just beyond my grasp. I told Jon to hold on tight to the rope and that I would make a reach for it. My only choice was to use one hand on my ice axe to pull myself up higher, over the top of the ice axe, and then try to reach for the rope with the other. Resetting the ice axe any higher into the blue ice was not an option, because if I pulled it out of the ice and tried to reset it, I would instantly slide down the chute. I had to hope that the tip of my ice axe would hold in the blue ice well enough for me to put all my weight on it.

With all my strength, I pulled myself up on the ice axe and made a daring reach for the rope. Luckily, the ice axe held and I managed to grab the rope. I tied off the rope to my climbing harness with a figure-eight knot; then using one hand on the rope and the other on my ice axe, managed to work my way up the chute to the ridge to join Jon.

The ridge, like the peak, was covered in spring snow, so it was a fairly easy walk from there up to the peak. Once we had ascended the peak, which was fully engulfed in a layer of snow, like an overfilled ice cream cone, we had to be diligent in finding the exact highest point. There was a real danger on the top, if we weren't careful, of stepping on an overhanging cornice and having it break off. Eventually, though, we settled on the highest point, changed into our warmer wool clothes, and finally had a chance to look around.

The views were stunning; looking westward over Revelstoke toward the Monashees we could see the very distant Pacific Coastal Range; and looking southward we could see the beautiful Purcells and Valhalla Ranges, in the south-eastern Kootenays; a perspective and vista I had never seen before.

Also from there, we could see Ghost Peak, a mountain that can only be seen from certain angles yet is so obvious once you see it, tucked in behind Mount Mackenzie and Cartier Mountain.

We grabbed a bite to eat and something to drink. We were past our turn-around time of one p.m. and knew getting down would be just as hard and take just as long as getting up.

Jon looked over the edge of the peak and suggested that it would be easier to belay ourselves off the summit, rather than climb down. Roping up to belay down takes just as much time as it does to climb, but in this situation I agreed with Jon. So he belayed me off the peak, down to a ledge below and then rappelled down after me, leaving behind one of his climbing straps tied into the rocks at the peak—homage to the mountain.

We worked our way slowly down the face of the mountain, following our route up, until we came to the ledge above the glacier we had first stood on together. Jon wanted to belay and rappel again off the ledge; I remembered the route we had climbed up and wanted to climb down that way. I felt our rope was not long enough to reach any meaningful advantage below and the time was running out. I did not want to be stuck on the mountain in the dark.

We stood on the ledge discussing our options for longer than we should have. The warm spring weather throughout the day had melted away some of the winter snow, and throughout the climb we had felt drops of water and pieces of ice and snow falling on us from up above.

Then all of a sudden we heard a terrifying rumble, and, looking up, Jon yelled, "Oh no, rocks, rocks! Get down."

I looked up from the ledge and saw an avalanche of rocks coming down from the top of the mountain. They had broken loose under the warm spring sun's thaw and were coming down right on top of us. Some of the rocks were the size of watermelons;

it was like a truck from up above had dumped a load of rocks over the edge on top of us.

I was standing in the widest part of the ledge; Jon, on the other hand, had removed his pack and was standing on the narrowest part. Neither of us had helmets. Jon crouched down on the ledge sideways, with one leg hanging over the edge. I turned towards the rock, placed my hands over the top of my head, leaned into the wall and closed my eyes, still with my pack on.

My life flashed before me, I thought that either I would be hit by one of the rocks and killed, or hit by one of the rocks and knocked off the ledge to my death. Everything suddenly turned into slow motion.

The sound was horrible from the rumble of the rocks coming down. The noise turned into bullet-like sounds as the bigger rocks bounced off the mountain, making gunshot-like "*pings*" all around and above us. I could smell the rocks hitting the mountain, like the fresh smell of gunpowder after you fire a rifle.

On and on the rocks fell, and then exactly as I had been fearfully dreading, one hit me. I felt a searing pain as a rock glanced off my right shoulder; fortunately I was partially protected by my pack. I managed to hang on and stay on the ledge.

Then I heard Jon yell out in pain, and then silence. I couldn't look over. I had my eyes closed. I didn't know if he was still on the ledge or not. Eventually, after what seemed like an eternity in hell, the sound of the rocks subsided, and dirt and gravel came pouring down on me. I stood there on the ledge, eyes still closed, hands still over my head, leaning into the mountain, getting covered in dirt.

When it finally stopped, I shook my head and looked over to where Jon had been crouched down, he was still on the ledge.

I said to him, "Are you all right?"

He said one rock had hit him square in the back. I told him one had hit me on my shoulders. All I wanted to do at that point was get off that rotten pile of rocks.

I told Jon, "I'm climbing down the way we came up!"

I worked my way down off the ledge and eventually came to the edge of the glacier. I found the nearest vantage point to get onto the glacier, then with a mighty leap, I jumped from the rock as far as I could onto the glacier, making sure I cleared the moat and the brittle edge. When I hit the glacier I slid down to a shallow point in the glacier, a bergschrund, and came to a stop. I was finally safe from the wretched mountain.

I looked back up the mountain and saw Jon working his way down. He was hurt, it took him a long time to get down to the edge of the glacier. When he finally reached the glacier, he yelled down at me, "How did you get onto the glacier from here?"

I yelled back, "I jumped. Throw me your pack and then jump out as far as you can."

Jon threw his pack and it slid down to where I was standing. Then he jumped down and landed safely on the glacier. When he approached me I ran up to him and hugged him, "Thank God you are all right," I said. "I love you, man!"

We both just stood there in a trance-like silence, looking back up at the wretched mountain and thanking our lucky stars we had both survived the climb.

"You know what, Jon?" I said, finally breaking the silence. "I know now I can climb anywhere or anything if I really have to. But I don't really have to. So I am never climbing again."

And to this day, I have never climbed again.

EPILOGUE

Several months after the passing of my father, I received a letter from my estranged aunt, one of my father's sisters. She didn't say much, not even that she was sorry to hear about his death. She just said that she had received a letter from my father a few months before his passing and wanted me to have a copy of it.

I've only ever read the letter a few times since, and only to a select few special people, never once without tears, most of the time with outright sobbing while I read it. Perhaps it helps me grieve, perhaps not. With tear-filled eyes as I write, this was his letter:

What Is the Purpose of Man?

What is the purpose of man, especially in this day and age? One of our Christian Fathers a few hundred years ago said, "Man is here to love God and enjoy Him forever."

I believe Saint Anselm was on the right track when he said that. I can relate to that, as a parent with children and grandchildren; a parent wants what is best for them, and if possible to take their problems and ease their burdens so that they can enjoy life without financial worries or emotional problems. Realistically, parents cannot do that. Their children and their grandchildren will get into all sorts of crises in their lives. Sometimes parents will be able to help, sometimes all they can do is wait and love them, and vicariously suffer with them through their troubles. What then does God do? There are over five billion people on this spaceship earth—is that too many, or not enough? Is it possible He has a plan and purpose for each and every one?

Sitting here in Revelstoke one looks in wonder at the mountains, with their glaciers and robes of dark green forests. Millions, no, billions, nay, trillions upon trillions are the snowflakes that make these glaciers. More snowflakes than can be imagined or counted to make a glacier—and there are thousands of glaciers. Can one tiny snowflake make a difference?

And trees, they grow in endless profusion, stretching as far as the eye can see and beyond. Does one tree really matter?

We had a mountain shower the other day. It rained for about half an hour, and while it rained the sun shone, and there was a rainbow, and as the rain lessened, the colours faded, only to brighten again as the rain increased. How many millions or billions or trillions of raindrops does it take to form a rainbow? For a second, perhaps only a split second, a raindrop falls to act as a prism to array in thin air all the colours of the rainbow—why? In this day and age of logic and scientific research and reason, there's no reason—i.e., purpose—given for a rainbow. Is there any other explanation than it is there to give beauty to delight the eyes of man?

So too I love what the prophet Isaiah says in his 56th chapter. "For as the heavens are higher than the earth, so my ways are higher than yours, and my thoughts than yours. As the rain and snow come down from heaven and stay upon the ground to water the earth, and *cause the grain to grow and produce seed for the farmer and bread for the hungry, so also is my WORD. I send it out and it produces fruit. It shall accomplish all I want it to, and prosper everywhere I send it. You will live in joy and peace. The mountains and hills, the trees of the field, all the world around you, will rejoice. Where once were thorns, fir trees will grow; where briars grow, myrtle trees will sprout up. This miracle will make the Lord's name very great, and be an everlasting sign of God's power and love."*

A single raindrop falling from the sky passes, for a split second of our time, through a sunbeam and becomes the catalyst to display the colours of the rainbow. Surely it is possible that we, passing through this world, our life a split second in God's time, may be a catalyst to show forth the brilliance, the radiance, the glory of God; or if we are not dazzling and no light shines through, we can, in our own way, be like that little drop of water, which nourishes the seed for the farmer so the hungry can be fed.

I am old, no longer does my family rely on me. When my children are around they now drive the car. I am not expected to organize birthdays or anniversary events, my children do. Out of courtesy, they allow me to cook the steaks at a barbeque. A titular head of the house, a groundskeeper and maintenance man. I feel like that little snowflake that has turned to an ice crystal in the glacier, waiting an eternity to come into its own when it will emerge on the edge of the glacier and be melted by a warm summer's sun and go rushing down the mountainside to give drink to some thirsty flower, or flow in a stream to a river and thence to the sea, the great Nirvana of all raindrops. . . .

The End